2024 THE FOURTH INDUSTRIAL REVOLUTION

STEPHEN HAAG
DARIAN HAAG
TREVOR HAAG

Janus Press, LLC

Janus Press, LLC

Printed in the United States of America
First Printing: April 2024

Library of Congress
ISBN: 979-8-9857439-2-0

Cover design by Mykhailo Uvarov

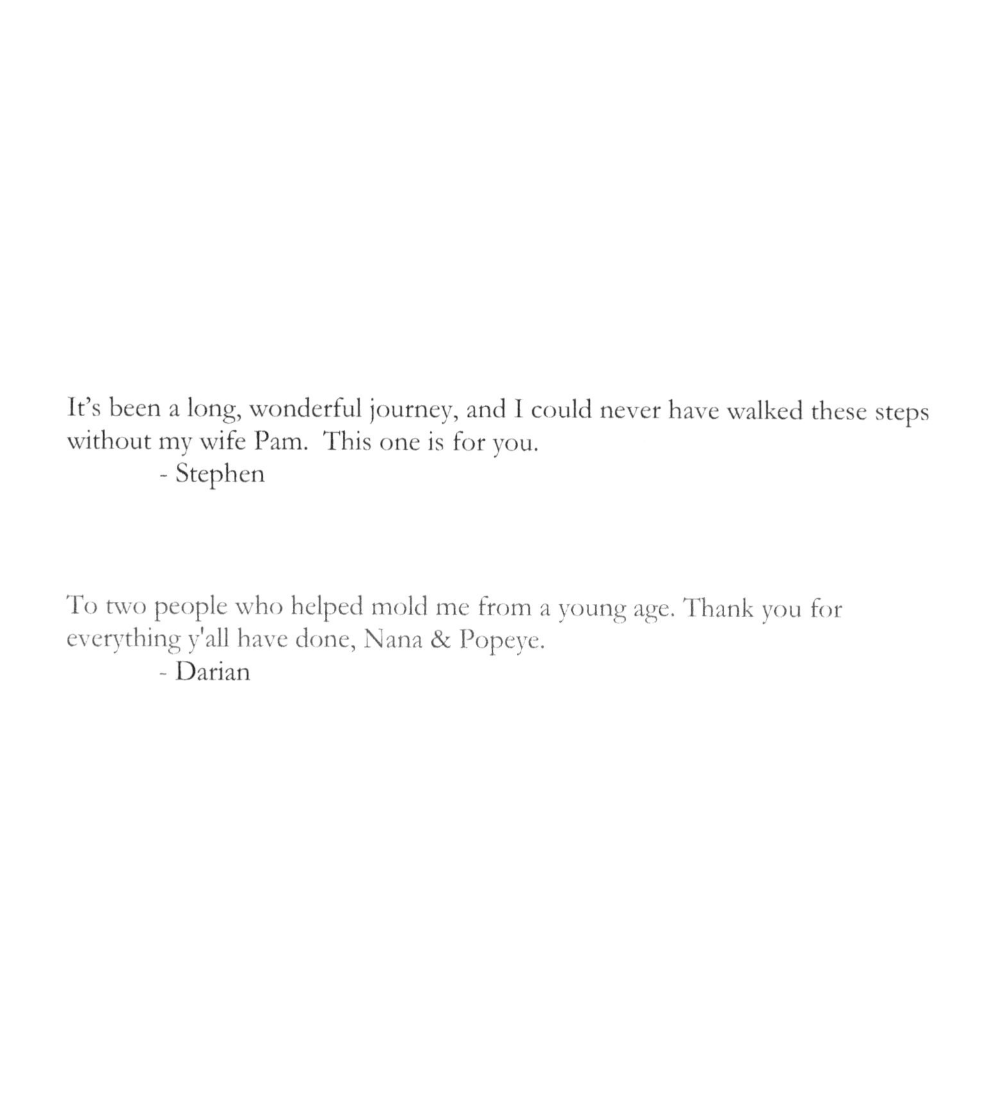

It's been a long, wonderful journey, and I could never have walked these steps without my wife Pam. This one is for you.

- Stephen

To two people who helped mold me from a young age. Thank you for everything y'all have done, Nana & Popeye.

- Darian

Table of Contents

THE SETUP 1

Past, Present, and Future

So Many Possibilities

CHAPTER 1 13

The Internet of (All) Things

A Water Bottle that Can Diagnose the Flu

CHAPTER 2 27

Blockchain and Cryptocurrency

$250 Million for a Pizza

CHAPTER 3 51

Artificial Intelligence

A Cat Mistaken for Guacamole

CHAPTER 4 75

Extended Reality

No Matter Where You Don't Go, There You Are

CHAPTER 5 91

3D Printing

One Pair of Shoes for Life

CHAPTER 6 107

Autonomous Vehicles

The Death of the Driver's License

CHAPTER 7 121

Drones

Male Bee, Bagpipe, Bladder Fiddle, or Flying Vehicle?

CHAPTER 8 131

Energy Technologies

Harvesting and Storage

CHAPTER 9 147

Sensing Technologies

Hearing, Seeing, Feeling, Smelling

CHAPTER 10 161

Communications Technologies

Anywhere Is the New *Where*

CHAPTER 11 169

Infrastructure Technologies

Quantum, Cloud, and Nano

About the Authors 183

Acknowledgements 185

Notes 187

Index/Definitions 197

THE SETUP

Past, Present, and Future

So Many Possibilities

This book is about the future and the present, but it's also about the past. Technologies like blockchain, autonomous vehicles, and artificial intelligence (AI) are very much a part of our future. They're also creeping into our present on a daily basis.

The past… well, that's a different story, in so many ways. To understand where we are today and the direction we're headed tomorrow, it often helps to know from where we started. That is to say, the past informs and shapes the present and the future.

It's also quite interesting to read quotes from the past about the future.

\\\ GREAT QUOTES FROM THE /// PAST ABOUT THE FUTURE

Let's call these *notable quotables.*

"Rail travel at high speed is not possible, because passengers, at high speed, would die of asphyxia."

1823, Dr. Dionysius Lardner, Professor of Natural Philosophy and Astronomy.

"X-rays will prove to be a hoax."

1883, Lord Kelvin, President of the Royal Society.

"Everything that can be invented has been invented."

1899, Charles H. Duell, U.S. Patent Office Commissioner. (Since 1900, there have been more than 11 million patents issued. Sorry Chuck.)

"Man won't fly for a million years - To build a flying machine would require the combined efforts of mathematicians and mechanics for 1 to 10 million years."

December 8, 1903, *The New York Times.* (You have to love this one. 9 days later 2 brothers, Orville and Wilbur Wright, made history with the first successful flight at Kitty Hawk.)

"The horse is here to stay but the automobile is a novelty - a fad."

1903, The President of the Michigan State Savings Bank advising Henry Ford's lawyer not to invest in The Ford Motor Company.

"If I had asked people what they wanted, they would have said faster horses."

Date unknown, but approximately the early 20th century, Henry Ford.

"A rocket will never be able to leave the Earth's atmosphere."

1936, *The New York Times.* (Tough times for *The New York Times.* It makes the notable quotable list twice.)

"I think there is a world market for maybe 5 computers."

1945, Thomas Watson, CEO of IBM.

"Television won't last long because people will soon get tired of staring at a plywood box every night."

1946, Darryl Zanuck, movie producer, 20th Century Fox.

"If excessive smoking actually plays a role in the production of lung cancer, it seems to be a minor one."

1954, W.C. Heuper, National Cancer Institute.

"We don't like their sound and guitar music is on the way out."

1962, Decca Recording executives while rejecting The Beatles.

"There is no reason anyone would want a computer in their home."

1977, Ken Olson, President of Digital Equipment Corporation.

"I predict the Internet will soon go spectacularly supernova and in 1996 catastrophically collapse."

1995, Robert Metcalfe, Co-Founder of 3Com.

"Children just aren't interested in Witches and Wizards anymore."

1996, Anonymous publishing executive writing to J.K. Rowling.

"There's no chance that the iPhone is going to get any significant market share. No chance."

2007, Steve Ballmer, CEO of Microsoft.

Looking back into history, it's hard to imagine that (intelligent) people would have made such statements. But those types of statements are still being made today. Consider this quote appearing in *The USA Today* (March 16, 2023), "I see no potential for it [AI, specifically ChatGPT] in medicine. By their very design, these large-language technologies are inappropriate sources of medical information." In 2022, when we wrote the first edition of this book, ChatGPT and other large-language AI models didn't even make the radar, so to speak. Now, we're debating their future and role in various industries like entertainment and medicine.

And here's one for cryptocurrency from 2018, "In terms of cryptocurrencies generally, I can say almost with certainty that they will come to a bad ending. If I could buy a 5-year put on every one of the cryptocurrencies, I'd be glad to do it…" (A "put" is a bet that the value of something - in this case, cryptocurrency- will go down. If the value does go down, you make money. If the value goes up, you lose money.) That quote comes from Warren Buffet, without a doubt one of the most famous and successful investors and money managers of all time. Who knows, he may very well be the one person to be right about the future. Then again…

\\\ THE INDUSTRIAL REVOLUTION /// JOURNEY

Indeed, we're with you today to journey into the future, the 4th industrial revolution, which in many respects is already here today in the present. Although, we would contend that what we've seen of the 4th industrial revolution thus far is just the tip of the iceberg in terms of the technologies and their innovative uses.

Time for a history lesson. Don't worry, it's fast. You will get to see 275 years in one image.

Agrarian Age

Prior to *industrialization*, we lived in what you might call the *agrarian age*, a time when land was power. Generally, the more land you owned the more powerful you were, and not too many people were actual landowners. We use the term "industry" to refer to cottage industries in the agrarian age, people offering a skilled trade from their homes while still working the land on which they lived. A blacksmith, for example, forged metal products like shoes for horses from their home to provide supplemental income but had to rely greatly on also working the land around the home for food.

Machinery, if you want to call it that, was powered by humans and animals.

Inventory, as we know it today was almost non-existent. Instead, we had bespoke products. If you wanted a pair of new shoes, you didn't go into the shoemaker's store and try on different styles, colors, and sizes. Instead, the shoemaker took your measurements and made custom shoes to fit you; originally the term *bespoke* meant something made specially for you. Not too many people could afford bespoke products; they wore hand-me-downs and/or made their own.

It truly was a time when people lived on the land and off the land. (Small play on words.)

Average life expectancy in the U.S.:

- Men - early to mid 30s
- Women - slightly more by a couple of years

The notion of retirement and living out your golden years didn't exist, especially the "golden" part.

Life was tough back then, everyone. Toilet paper wasn't invented until the 1850s. (Yuck.) Toothbrushes weren't invented until the late 1700s. The first bar of soap wasn't sold in a retail store until 1879. (Stinky.) Life was short and generally tough, except for the very wealthy.

1st Industrial Revolution

The 1st industrial revolution began in the 1750s-ish and lasted until the 1840s-ish. Historians love to argue the exact time frame, but it's generally accepted to have been from the mid 18th century to the mid 19th century.

One of the major catalysts for the 1st industrial revolution was the use of water and steam to power mechanized factories. The textile industry, the dominant industry at the time, led the way with water- and steam-powered cotton, wool, and linen spinning machines, power looms, and the cotton gin (separation of the seed from the cotton).

In this regard, we went from people/animal power to water/steam power. This led to unprecedented increases in worker productivity, often as much as fifty-fold for daily output.

Other important advancements during this period included chemical manufacturing, iron production, the steam engine locomotive, cement, glass making, and gas lighting.

People began moving off their small farms (which they didn't own) to the cities to get jobs in the factories. The notion of employment, as we know it today, emerged. Large industries (textiles, manufacturing, mining, transportation, etc.) with multiple large competing firms began to form.

Average life expectancy in the U.S.:
- Men - 41
- Women - 46

2nd Industrial Revolution
The 2nd industrial revolution began in the 1870s-ish and lasted until the beginning of World War I, around 1914.

The big catalysts for the 2nd industrial revolution were electricity, the formalization of power grids, the internal combustion engine, the telegraph, and the telephone. So, once again, we shifted the basis of mechanized power, from water and steam to electricity and the gas-powered internal combustion engine. This led to bigger, better, and faster mechanized factories and again to tremendous increases in worker productivity.

The telephone and telegraph enabled communications over great distances in a very short period of time. We used power grids to provide electrification to cities. Iron and steel production processes became much better. Paper-making processes greatly improved, leading to advancements in general education (basic reading and writing). The use of ammonia in fertilizer led to unbelievable crop yields.

More and more people moved from rural farms to find employment in cities, which cropped up around every major manufacturing operation.

Average life expectancy in the U.S.:
- Men - 68
- Women - 70

3rd Industrial Revolution
The 3rd industrial revolution began in the 1950s-ish. It's safe to say we're still in the latter stages of the 3rd industrial revolution but rapidly moving into the 4th industrial revolution.

This is the age in which you've lived your entire life, punctuated by computing and digital technologies, communications technologies, the Internet, smartphones, and so on. It's often referred to as the ***digital revolution***, the ***digital age***, and the ***information age***. (You are referred to as ***digital natives***, while people in previous generations are referred to as ***digital immigrants***.)

In the 2nd industrial revolution, the biggest and most well-known companies in the world were those that could produce/sell the most physical goods. They included the likes of Standard Oil, General Motors, Ford, US Steel, Sears/Roebuck, General Electric, Bethlehem Steel, International Harvester, Borden, and Goodyear Tire & Rubber.

As we near the end of the 3rd industrial revolution, today is quite different. The biggest companies in the world (using early 2024 market capitalization or market cap) are:

1. Apple
2. Microsoft
3. Saudi Aramco (the only remaining oil company in the top 10)
4. Alphabet (Google)
5. Amazon
6. NVIDIA
7. Meta Platforms (Facebook)
8. Berkshire Hathaway (investment company led by Warren Buffet)
9. Tesla

Seven of the 9 largest companies in the world are all tech companies. (To be sure, Amazon and Tesla are <u>tech</u> companies. Amazon's most profitable business segment is Amazon Web Services. Not familiar with AWS… read the Cloud Computing section in Chapter 10 on Infrastructure Technologies.)

Within the 3rd industrial revolution, we had some interesting sub-revolutions:

- 3.1 Personal Computing, starting in the late 1970s… home computers, desktop computers, microcomputers, laptops, etc.
- 3.2 Electronic Commerce, starting in the 1990s… Amazon, eBay… literally any organization/business you can name
- 3.3 SoLoMo (Social-Local-Mobile), also starting in the 1990s and dominating the first 2 decades of the 21st century… TikTok, Snapchat, Instagram, Facebook, etc.

Average life expectancy in the U.S.:

- Men - 75
- Women - 80

4th industrial Revolution
And, finally, we have arrived where we are today, the 4th industrial revolution.

It will be interesting to read about the 4th industrial revolution 20 years from now. Did it start in 2015? Probably not. 2020, perhaps. 2030? We will most certainly be in it by then.

Regardless, the 4th industrial revolution is your present and your future. You and your generation are in control. Historians 20 years from now will write about how you and your generation shaped the 4th industrial revolution.

At the core of the 4th industrial revolution will be the following primary technology drivers (see Figure 0.1):

1. The Internet of (All) Things
2. Blockchain & Cryptocurrency
3. Artificial Intelligence
4. Extended Reality
5. 3D Printing
6. Autonomous Vehicles
7. Drones

Figure 0.1 **The 4th Industrial Revolution**

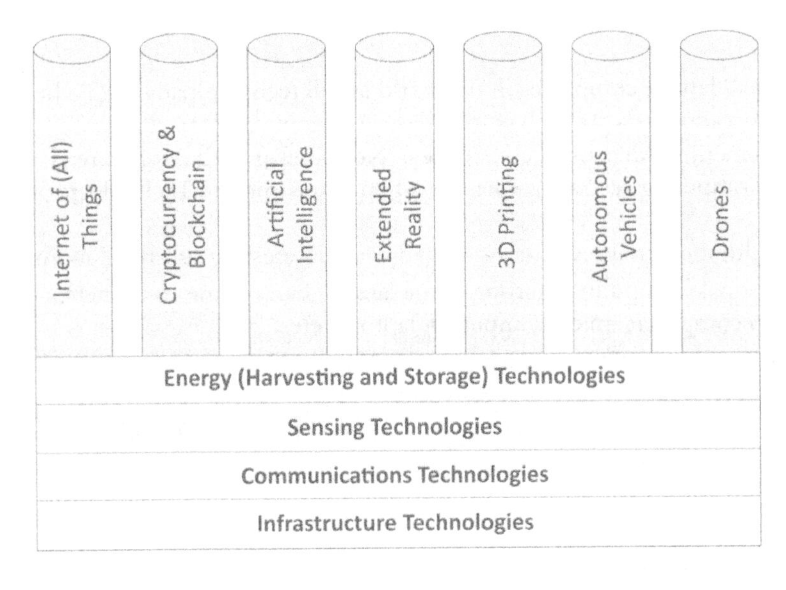

Those technologies will be supported and enabled by existing and advances in energy (harvesting and storage) technologies, sensing technologies (hearing, seeing, smelling, and feeling), and infrastructure technologies like quantum computing, cloud computing, and nanotechnologies.

And we'll see applications and advancements of the 7 core 4th industrial revolution technologies in/with things like geo-engineering, bio-technology, neuro-technology, digital twins, and much more.

Most importantly, all those technologies are the foundation on which you will build unbelievable innovations. Those innovations will touch, impact, and disrupt every aspect of daily life.

\\\ THE S-CURVE OF AN /// INDUSTRIAL REVOLUTION

You can think of each industrial revolution as an s-curve (See Figure 0.2).

Figure 0.2 **Industrial Revolution S-Curve**

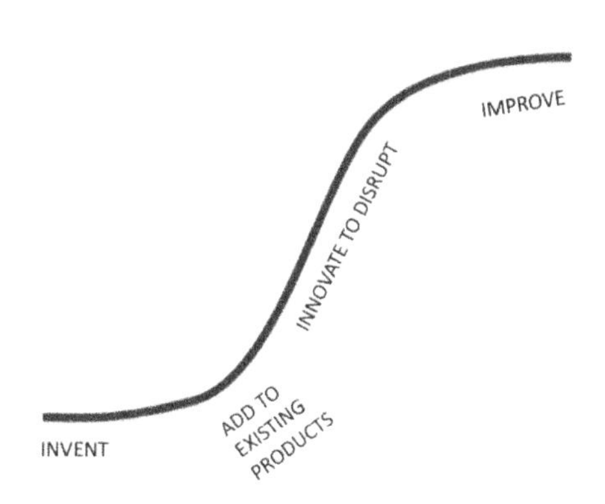

(Clayton Christensen was the first to apply the s-curve in the invention/innovation spaces. We highly recommend reading his works.)

It all starts with a bunch of new technological inventions, like electricity, the internal combustion engine, telephone, and telegraph of the 2nd industrial revolution. During the invention stage, market acceptance is low, as the technologies are unproven, expensive, perhaps slow and clunky, and not widely available. At some inflection point, the rise over the run changes dramatically as the technologies get better, faster, cheaper, and more reliable, and as people start to innovatively figure out how to use the technological innovations to reshape existing products and services.

Market acceptance then skyrockets, the technologies continue to get better, faster, and cheaper, and society in general becomes more "comfortable" with new ways of doing things, not to mention that businesses figure out how to make money with their innovations. New products and services emerge based on the new technologies that weren't possible with previous technologies, leading to dramatic industry disruptions.

During the latter part of an industrial revolution, the speed and pace of innovation slows and people begin to focus on creating effectiveness (doing the right things) and efficiency (doing things right… in the least amount of time, at the lowest cost, etc.).

And then - suddenly - a bunch of new technological inventions emerge, and the birth of a new industrial revolution begins.

Figure 0.3 captures the previous 275 or so years and our progression through the first 3 industrial revolutions and the beginning of the 4th.

Figure 0.3 **275 Years in One Image**

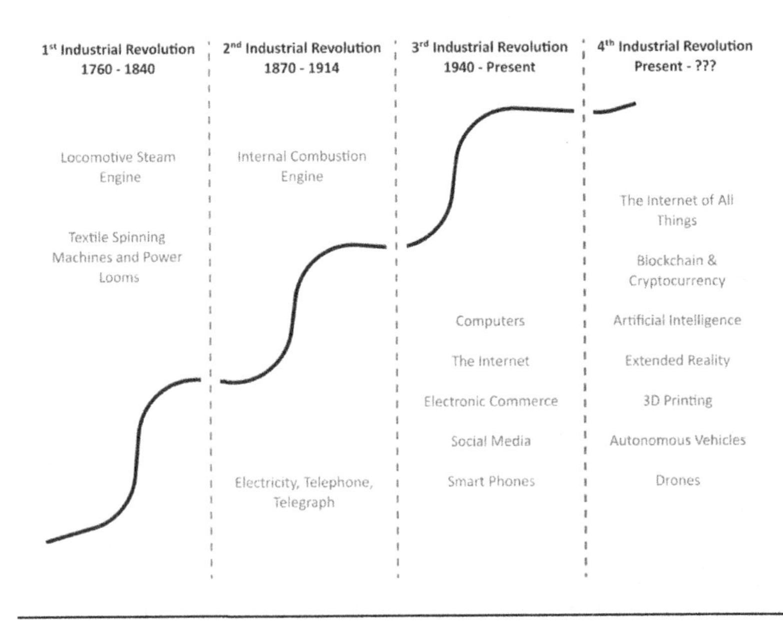

If you look to the upper right corner, you'll notice just the beginning of the 4th industrial revolution s-curve. Fortunately for you, we are just at the beginning of the 4th industrial revolution. For the most part, we haven't even begun to scratch the innovation surface using technologies such as AI, IoT, blockchain, 3D printing, and much more.

We believe, without a doubt, that the next 20 years or so will be the most innovative time in the history of the world. Your opportunities are simply unimaginable.

Best of luck inventing the future.

As the 3rd and 4th industrial revolutions have advanced technology - specifically computing technology - as their basis, let's take a bit of a closer look at the advances within the technology itself.

Electronic computing as we know it began in the 1940s with the ENIAC (Electronic Numerical Integrator and Calculator), the world's first all-electronic computer. The ENIAC took up most of the floor of a large building in Philadelphia. If you wanted to work on the computer, you actually worked *inside* the computer, often using a ladder to reach certain parts. When someone turned on the ENIAC, the lights of the surrounding neighborhoods dimmed.

The first electronic computers were based on an infrastructure technology called vacuum tubes. From vacuum tubes, we innovated to transistors, then to integrated circuits, and then to VLSI (very-large scale integration) today. Of course, along the way those infrastructure technologies became faster, cheaper, and a lot smaller in size.

In 1965, Gordon Moore forecasted that the number of components on an integrated circuit (a rough representation of technology speed, capacity, and power) would double every year until it reached an amazing 65,000 by 1975. When that prediction was proven true in 1975, the yearly doubling concept of technology speed/capacity/power became known as Moore's Law.

Moore's Law obviously isn't a government law, nor is it a law of nature. Rather, Moore's Law is a statement of the trajectory of technology with which its speed, capacity, and power are doubling every year. (Along with that doubling of power, the technology itself is becoming roughly half as expensive.)

As computing technology becomes more powerful with costs also being cut in half, we can do more things and be more innovative with the technology. Take artificial intelligence, for example. AI takes a lot of computing horsepower, horsepower that didn't exist in the midst of the 3rd industrial revolution. It may have existed on a very limited basis, but it wasn't anywhere near powerful enough nor small enough to put in an automobile and have the automobile drive itself (i.e., autonomous vehicles).

The same is true for blockchain, a method of storing and managing information simultaneously on hundreds, if not thousands, of servers which could be located all over the world. Blockchain is only possible because of the capacity and speed of infrastructure technologies (for example, the communications technologies that make possible the high-speed transfer of information across the globe).

Now of course, you may be thinking, "Wait a minute… smartphones aren't getting cheaper or smaller. If anything, they're more expensive and larger than previous releases." Well, yes and no. Smartphones are getting larger and more expensive, but they also include much more technology, speed, capability, and capacity. The new models have 3 cameras, augmented reality, more storage capacity, better communications (5G, for example), and longer lasting and faster (wireless) charging batteries. Companies like Apple don't want to simply make successive iterations of smartphones smaller and cheaper. They want to make "better" smartphones with each release that have more functionality and capability, thus raising the price (which most of us seem completely willing to pay).

Okay, enough of a history lesson, let's launch into today and - more importantly - tomorrow.

CHAPTER 1

The Internet of (All) Things

A Water Bottle that Can Diagnose the Flu

QUESTION: Are there more stars in our universe or grains of sand throughout the world?

Scientists today estimate that there are 7.5 sextillion grains of sand, or 7,500,000,000,000,000,000. It's a big number. Our universe, however, contains an estimated 70 septillion stars, or 700,000,000,000,000,000,000,000. That means there are roughly 10,000 stars for every grain of sand.

So, you should get this question right: Are there more people in the world or more devices connected to the Internet? You guessed it… there are more devices connected to the Internet than the entire population of the planet by a factor of about 8 to 10, depending on when you read this.

As you can see in Figure 1.1 it was about 2007-ish when the number of devices connected to the Internet exceeded the human population. [1] In 2020 there were about 7 devices connected to the Internet for every person, and by 2030 Cisco estimates that there will be 500 billion devices connected to the Internet, about 60 devices per person assuming the world's population in 2030 is 8.5 billion. [2]

Figure 1.1 **World Population Versus Internet Connected Devices**

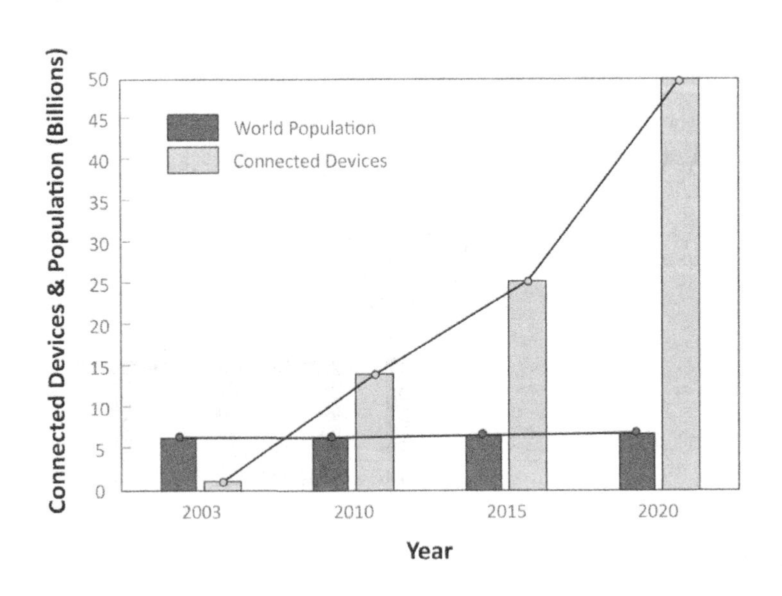

We aggregately refer to all those Internet-connected devices as the ***Internet of Things (IoT)***, formally, a network of Internet-connected objects that collect, process, and exchange data.

\\\ IOT IS ARRIVING /// IN STAGES

It's certainly true that IoT already exists in the present, but it's arriving in stages. We've got basic IoCT (stage #1), we're deep in the middle of IoET (stage #2), and we're just now moving into IoAT (stage #3). Let's talk in more depth about the evolution of IoT (See Figure 1.2).

Figure 1.2 **Internet of (All) Things Evolution**

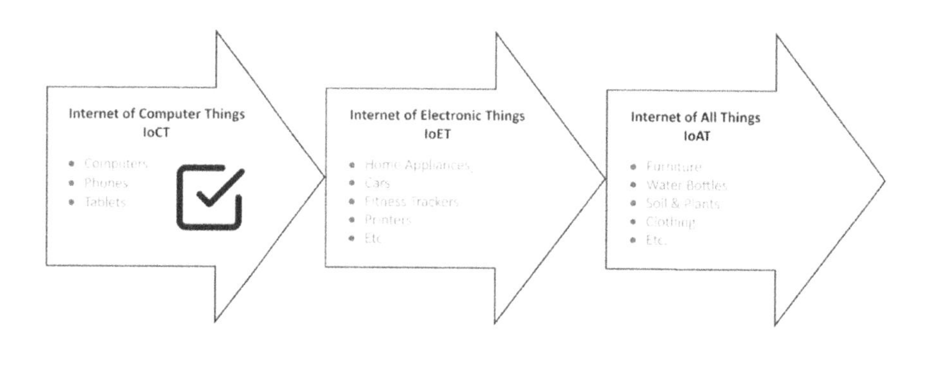

Stage #1: IoCT (Internet of Computer Things)

Stage #1, ***the Internet of Computer Things*** or ***IoCT***, began in the 3rd industrial revolution with all your computers and computer-related things connected to the Internet, and communicating with each other and sharing information, data files, photos, etc. You can use your phone and share your photos to your computer (AirDrop if you're in the Apple world, Nearby Share if you're in the Android world). You can use your laptop or tablet to locate your phone. You can easily and dynamically create, edit, and share files among your computer devices and other people using services like iCloud, Dropbox, and Google Docs & Drive.

Stage #2: IoET (Internet of Electronic Things)
We are deep in the middle of stage #2, ***the Internet of Electronic Things*** or ***IoET***, as we connect all electronic things to the Internet.

In this case, think about your household electronic devices such as a printer, washing machine, clothes dryer, refrigerator, heating and air conditioning, garage door, water sprinkler system, stereo, and TV, just to name a few.

A printer seems obvious, as it's a natural extension of your "computer system." But it's more than just wirelessly connecting your computer to your printer. You can subscribe to a printing service such as HP Instant Ink. If so, you connect your printer via the Internet to the service. With every print, your printer communicates back to HP the number of pages you printed, color versus black and white, and how much of each toner color you have used. When HP anticipates, based on your printing habits, that you'll soon be running low on ink, it automatically sends you the ink refills you need. (Of course, HP will charge your credit card for the cost of the ink refills.)

Google Nest is another example in the IoET space. It offers a wide range of Internet-connected home conveniences including Learning Thermostat, Doorbell, and Protect (smoke and carbon monoxide detection). Ring also offers a variety of IoT-based solutions for your home including indoor and outdoor security cameras and its signature video-based doorbell systems.

You can connect your phone, tablet, or computer to any of these types of systems to control your lights, open and close the garage door, turn on or off the water sprinklers, and much more. So, they fall into the IoT space, a network of Internet-connected objects that collect, process, and exchange data.

You may even be able to pair your phone with the clothes washer or dryer in a college dorm and receive a text message when your laundry is done. And you can most likely pay for use of the machine with your phone.

There are many other examples, many of which we consider in the ***smart home*** category (also called ***home automation*** or ***domotics***), building automation systems and intelligence for your personal home. You can use an app to control your lights, garage door, and setting for your water sprinkler system. And, of course, that whole genre of be-as-lazy-as-you-want technologies like Amazon Alexa and Google Assistant can answer questions, control various parts of your home, and generally make you feel like you never have to leave your couch (or touch a remote control).

Smart homes sound great, but there is the issue of interoperability. That is, will all your intelligent devices - Samsung, GE, Amazon, Microsoft, Apple, Google, etc. - "play nicely together" and share information without any connectivity issues? We hope the answer is yes through Matter. ***Matter*** is an open-source connectivity standard for smart homes and IoT devices. Matter was created in 2019 by the Connectivity Standards Alliance and now includes more than 200 organizations like Google, Apple, Amazon, and Microsoft.

There's lots of other IoET applications as well for commercial buildings, transportation systems, weather alert systems, and so on. Many of these fall in the category of ***smart cities***, technology-based city environments that use a wealth of technologies to gather data and use that data to more efficiently deliver services related to such things as utilities, transportation, crime detection, medical care, waste removal, and snow plowing. Smart cities systems that apply IoT to improve the delivery, efficiency, and reliability of electricity are called ***smart grids***.

Big cities like New York City, San Jose, and Boston already have elements of the smart cities concept in place. New York city has implemented smart initiatives for water control, waste management, navigation, and tourism. Boston has implemented a system to improve traffic efficiency. It even lets parents track their kids' school buses. And San Jose is using IoT sensors to monitor air pollution to address air quality issues.

Industrial Internet of Things

All sorts of new terms and acronyms are popping up everywhere in the 4th industrial revolution space. One of those is the industrial Internet of Things.

The ***industrial Internet of Things (IIoT)*** refers to the use of IoT in commercial and industrial settings as opposed to consumer-oriented IoT like Google Nest and Ring. Smart cities use of IoT falls into the category of IIoT.

Major manufacturing operations are using IIoT to monitor manufacturing equipment. In this case, IoT sensors that measure vibration, for example, are attached to manufacturing equipment to detect early warning signals that equipment may be coming out of alignment and need maintenance and/or repair.

Worldwide, the industrial Internet of Things market was estimated to be $394 billion in 2023. Projections for 2024 are approximately $467 billion. [3]

We're here in the middle of the innovation aspects of stage #2, with lots of interesting and innovative opportunities still to come.

Stage #3: IoAT (Internet of All Things)
We are just starting to dip our toes into stage #3, the ***Internet of All Things*** or ***IoAT***, the connection of all things non-electronic to the Internet.

Let's fast forward and envision a possibility of the Internet of All Things. You get up one morning feeling a little groggy. Perhaps you didn't sleep well but something seems to be crawling up the back of your neck. You get ready and head off to school.

At some point, you take a drink from your water bottle and shortly thereafter get a text message. It seems that your doctor's office received a communication from your water bottle that you may be coming down with the flu.

Seem impossible? Not really and it may very well happen in stage #3, the Internet of All Things. Gatorade's Smart Gx bottle is one such example. Using IoT sensors in the bottle, hydration levels of the user are collected and tracked, while calculating beverage intake requirements based on the user's profile and level of activity (stored in the Gx app). The technology provides reminders when to take another sip, when the bottle is empty, how many more ounces need to be consumed, and more. Add in a sensor that can analyze bacteria, and your smart water bottle now has health-related capabilities.

Home water sprinkler systems represent another IoAT opportunity. Irrigreen, for example, offers a water sprinkler system that uses a "seeing" water sprinkler head to automatically adjust its spray distance as it rotates its head. When the sprinkler sees that it is coming to a rock garden area, it reduces its spray length so as not to water the rocks. When spraying into a distant corner of the grass, the sprinkler sees that it needs to spray farther and adjusts its spray distance accordingly.

Water sprinkler systems can also use embedded ground sensors to determine when soil moisture is low and respond by increasing the watering. These same systems can connect with a weather app to respond appropriately to changing weather patterns. (It's going to rain heavily tonight, so no need to water the lawn this evening.)

Clothing is becoming Internet-connected. We've already put sensors in shoes to measure your speed and distance traveled. Research is occurring to literally weave sensors into the fabric of clothing. [4] Someday, you'll wash a load of clothes and your washing machine will alert you to the fact that you left your car key fob in your pants pocket and are just about to wash it (the key fob, that is).

We can now connect sensors to living things like trees to monitor growth, determine when watering needs to occur, and detect the presence of diseases. (Many tree and shrub diseases emit chemicals, which we can detect with IoT smelling sensors. Learn more about smelling sensors in Chapter 8 on Sensing Technologies.)

We're also putting sensors in furniture. Pillows and beds can monitor your sleep patterns. [5] Intelligent chairs can give you a slight vibration when you've sat too long in the same position.

We can even 3D print sensors directly on to human skin. [6] These types of sensors provide real-time biometric data to enable better healthcare.

We believe this is where you play, in stage #3 the Internet of All Things. We've just barely seen the tip of the iceberg in the IoAT space. Many, many wonderful innovations to come, just waiting for you and your free-thinking group of friends to bring them to market.

\\\ HOW IOT WORKS:
/// FROM SENSOR TO APPLICATION

An IoT system has 4 layers. Consider Waze, the popular navigational app. (Yes, Waze is an IoT application.) Seems a bit weird but read the table below from bottom to top.

IoT Layer	Waze
Layer #4: Application	Takes all that information, applies some algorithms and mapping functions, and shows you shortest routes, locations of traffic jams, locations of emergency vehicles, etc.
Layer #3: Data Management	Aggregates and organizes the information from potentially millions of current users by location.
Layer #2: Communication	Sends your information to the cloud via broadband cellular network (hopefully 5G).
Layer #1: Sensor	Uses your GPS to determine things like your speed, direction of travel, location, etc.

A few noteworthy observations.

Firstly, in the communication layer there are many options. If you're building an IoT application for use in or around your home, you'll most likely be using WiFi and perhaps Bluetooth, RFID, and/or NFC. Waze relies on broadcast cellular (5G, etc.) because it's what your phone uses when you're on the road. Again, many options here and we would encourage you to read Chapter 9 on Communications Technologies, especially if you're unfamiliar with things like RFID and NFC.

Secondly, is the notion of security. Most IoT systems fall in the category of *lean tech*, meaning that the system (usually because of its size) has only the minimum software and hardware necessary for it to execute its specific function or task. Unfortunately, one of the first pieces of software to be omitted is security. The thinking is simple: You don't need security software to run your water sprinklers based on soil moisture level.

But the absence of security software can lead to hacks and breaches. In 2013 for example, hackers penetrated an HVAC company that wirelessly monitored retailer Target's environmental control systems. Using that wireless gateway to the heating and air conditioning systems, the hackers were able to access millions of customer records on Target's large computers. [7]

Of course, you may think that hacking into your water sprinkler system won't do anyone any good. Think again. If someone can get into that IoT system, then they made their way into your home's WiFi. That's trouble.

\\\ THERE'S A SENSOR /// FOR EVERYTHING

A ***sensor*** is a device that monitors and measures the physical aspects of an environment. So, a sensor provides data about a thing. That *thing* could be just about *anything*… your location, someone walking into a room, the temperature, the presence of smoke, the pH level in water, etc.

There are, quite literally, hundreds of different types of sensors used in IoT applications, way too many to completely cover here. The list below provides general groupings of sensors.

- GPS (location), accelerometer (speed), gyroscope (direction)
- Proximity/ultrasonic (how close is an object), motion, infrared, tracking (edge detection and line following), vibration
- Barometric pressure, gas pressure, weight (pressure)
- Temperature and humidity
- Water level, water flow, moisture
- Magnetic detection and metal touch
- Chemical, gas, and carbon monoxide/dioxide
- Smoke, flames
- Light
- Sound
- Water quality (e.g., pH, chlorine-related)
- Image recognition, color recognition
- Biometric (includes voice recognition, fingerprint recognition, facial recognition, and occupancy counting… how many people are in a room).

Like we said, hundreds and hundreds of sensors. If you need to detect and measure something, there's an IoT sensor for it. Costs range from about $1 to $50 and beyond. For example, you can buy a nice AI camera for about $50. It does image and facial recognition. For less than $20, you can buy a fingerprint recognition sensor that can store, recognize, and validate up to 240 fingerprints.

Natural or "Sensed" Data

You should start to think about the whole IoT space in terms of *natural* or *sensed* data, captured data in its natural, raw form at the point of origin. You could count the number of people in a room, or you could use an IoT sensor that "senses" when a person walks into the room. Add 1. It could also sense when a person leaves the room. Subtract 1. That's very easy to do with IoT.

You can build a toilet-base wrap-around IoT application that detects water (moisture) when the toilet overflows. That would be sensed data (water), and the system could then send you a text message.

You can buy moisture sensors for your backyard that turn on the water sprinklers when the soil moisture drops below a certain level.

You can strap a GPS unit to a bicycle helmet so you can monitor the whereabouts of your children.

You can build an IoT application and put smelling sensors in plastic bowls to detect when food is going bad in your refrigerator.

You get the idea. Each of those is about sensing the data and can provide more timely information.

Organizations are quickly moving to take advantage of capturing natural or sensed data. Amazon with its retail stores, for example, is rolling out a cashier-less and self-checkout-less shopping experience. You just walk in, get what you want, and walk out. Amazon uses the Amazon Go app on your phone, camera technologies with image recognition, and IoT sensors to capture sensed data about you (when you enter the store), what products you select (using cameras and IoT sensors), and your Amazon account (stored on the Amazon go app) to charge you for the products you select.

Capturing natural or sensed data yields so much more information than typical transaction data. Think about grocery shopping. When you check out, all your items get scanned. That's transaction data. But the grocery store doesn't know which aisles you chose or how you moved through the store, in what order you selected your groceries, or what you pulled off the shelf, chose not to buy, and placed back on the shelf. That sort of sensed data would be invaluable for a grocery store.

You should think critically about the Amazon example. Any retailer using such a system will know how much time you spent in the store, the order in which you traversed through the store and selected products, what products you looked at but chose not to buy, and so on.

We can capture natural or sensed data on the web very easily. Think about Netflix. It captures which movies you look at but choose not to watch. It captures movies for which you watch the previews but choose not to watch. That type of information is important to Netflix as it builds a profile of you and your movie habits.

Knowing what people choose not to buy is just as important - if not more important - than knowing what people do buy.

Tile, SmartTag, and AirTag as Examples of IoT Applications

We most often think of these as key finders because keys are the most often lost item in the home. But you can attach Tile-like devices to just about anything… remote controls, wallets, gloves, shoes, etc.

The better ones on the market include:

- Tile (tile)
- SmartTag (Samsung Galaxy)
- AirTag (Apple)
- Chipolo (Chipolo)

As an IoT application, each of these includes some sort of short-range communications technology to "find" or locate what you've lost. You can easily build your own version by incorporating something like Bluetooth.

\\\ EDGE COMPUTING /// AND IOT

Edge computing is a distributed computing model that moves the storing and processing of data closer to where the data is captured or sensed, i.e., the point of origin. So, IoT can be an implementation of the edge computing model. The notion is to shorten the distance - both physical distance and time - between layer #1 and layers #3 and #4. Refer back to the table on page 20.

In referencing back to that table, think about Waze. Waze is definitely not about edge computing. When data is captured about your driving (location, speed, etc.), it is sent to the cloud where that data is stored, organized, and processed. The results are then sent back to your device in the form of maps, driving recommendations, etc. So, the distance between the capturing of the data and the storing and processing of the data is significant. However, in order to shorten this distance, Waze would need sensors in every car and every street, not necessarily viable. While the edge computing model should be prioritized with IoT devices, it's not always achievable.

Now, let's think about an IoT application in your classroom. It uses facial recognition technology to take attendance for the instructor. (No more calling the roll.) As the data is captured, it can be processed right there in the classroom. The data doesn't have to be sent anywhere for storing and processing.

Okay, now that you know the basics of IoT, let's talk about the future.

\\\ THE END GAME: /// WHAT THE FUTURE MIGHT HOLD

Everything Will Be Connected

With the continued miniaturization of technology, increases in functionality, and decreases in price, it's hard to imagine what won't be connected to the Internet via IoT.

Water bottles, backpacks, bicycles, clothing, furniture… those all begin to make obvious sense and the data we can sense from them can be used in many innovative ways.

Rugs in your home, a door, a doorstop, your lawnmower, even your toothbrush… at first glance those seem a bit less obvious for an IoT application. But simply stop and think innovatively. A toothbrush that can detect cavities and diseases, monitor brushing effectiveness, tell you when to spit. We'll see them someday. (Oops, Oral-B already has one and it includes artificial intelligence. [8])

Security Will Be a Huge Concern

We cannot stress enough how important security is as a consideration in the IoT space. People have already proven that they can take control of an autonomous vehicle through hacking. [9] To be sure, IoT is an integral technology for autonomous vehicles. We told you about the Target hack that occurred via the HVAC IoT system.

Even in the development of lean tech, we cannot sacrifice security.

The Death of the House Key

With everything connected and our ability to sense things with technology, we will see significant shifts in physical security toward the use of biometrics. Think about security in terms of:

1. What you have, a key or an identification card of some sort such as your school ID.
2. What you know, a password.
3. Who you are, a unique biometric of you such as your fingerprint.

The first two levels are easy to steal and duplicate; the third not so much. We already have the use of biometrics on phones, computers, and other personal devices, and, to a certain extent, automobiles.

That trend will certainly continue and start to include things like access to your gym locker, bicycle lock, dorm building, apartment, and home.

You Don't Make a Pig Fatter Just by Weighing It

Good words of wisdom from the farm.

Gathering data does you no good if you don't use that data to take action or make better decisions.

If you want your IoT innovation to be successful, you have to find value in the sensed data.

Add Just a Pinch of…

So, how do you get started innovating in the IoT space?

If you recall our s-curve discussion, the application of new inventions starts by adding those inventions to existing products and services. It's an innovation technique we call, "Add Just a Pinch of…"

Take any existing product - a toothbrush, a water bottle, whatever. Now, simply add some IoT to it. What do you need to *sense*? What IoT sensors are required? What sensed data would you capture? How would you use that data? Now, build it. If you do so for a toothbrush for example, you have to understand that your minimum viable product (MVP) will be ugly, ugly. You'll probably have to hold the technology in one hand while you brush with the other. But the point is to prove that the technology works, regardless of how ugly it may be. That is called a *proof-of-concept prototype*. It doesn't have to be pretty; it just has to work.

Try adding just a pinch of IoT to the following products. For each, identify what data you will sense and why, i.e., don't just weigh the pig.

- School backpack
- Running shoes
- Tennis ball
- Rain gutter
- Mountain bike
- Dog food and water bowl
- Toilet (there's an interesting one)

As an alternative, you can focus on a specific type of IoT sensor and start to imagine possibilities of when/where it could be used. Take, for example, a simple pressure sensor that measures the presence of something by noting a change in weight. Can you envision possibilities for when this would be helpful for beds, car safety seats for children, lawn furniture, bookcases, etc.?

As a final alternative, pick a world problem/issue that you're passionate about. How can you use IoT to address that problem?

CHAPTER 2

Blockchain & Cryptocurrency

$250 Million for a Pizza

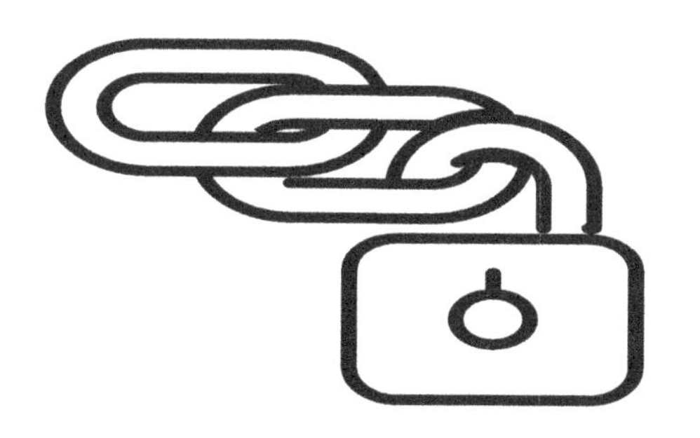

Time for another question: In 2010, someone bought the 2 most expensive pizzas ever sold. How much did those pizzas cost?

A. $1,000
B. $10,000
C. $100,000
D. $10,000,000
E. Upwards of a half-billion dollars

Sad story, as it turns out. (BTW, the answer is E.)

In 2010, Laszlo Hanyecz had accumulated 10,000 coins of something called Bitcoin, a fanciful outlandish digital version of currency that wasn't backed by any government. At the time, a single bitcoin was worth about $0.004, not even a penny. So, Laszlo spent the equivalent of about $40 in 2010 on 2 pizzas. $40 for a couple of pizzas isn't outrageous by any stretch of the imagination. But, in today's dollars 10,000 bitcoin is worth ever so slightly more than $40. (chuckle)

What's the value of 10,000 bitcoin today? Look up the value of a single bitcoin and quickly multiply that number by 10,000 and you'll get the exact amount Laszlo spent on 2 pizzas. (As we're writing this, a bitcoin is worth $42,000, so about $420,000,000 for 2 pizzas. Ouch.)

Of all the 4th industrial revolution technologies, cryptocurrency - along with blockchain - may very well be the most disruptive. No doubt, new industries, businesses, and ~~millionaires~~ billionaires will emerge in this space. Many traditional businesses and perhaps entire industries will not survive the disruption. Governments will have to rethink everything about monetary policy. You'll get to decide if you want your paycheck deposited in traditional currency or crypto.

Cryptocurrency and blockchain are as well perhaps the most dominant of the 4th industrial revolution technologies in terms of their coverage in the popular press. (Although, large language models, generative AI, prompt engineering, and other AI-related terms and technologies are also certainly garnering their share of the press these days.)

- In 2021, Aaron Rogers, the then-quarterback for the Green Bay Packers, decided to take part of his annual pay in Bitcoin and also give away $1 million in Bitcoin to fans. [1]
- In 2017, the Navajo Nation made a big push into mining Bitcoin. [2]
- In 2021, the newly elected mayor of New York City (Eric Adams) took his first three paychecks in Bitcoin. [3]

- In 2021, the city of Miami, FL announced it would distribute some of its Bitcoin yield to residents. [4]
- In 2021, Trevor Lawrence, the number 1 pick in the NFL draft, announced he would take his entire bonus (just over $22 million) in Bitcoin. [5]

We'll stop there, but you get the idea hopefully. You can do a quick search on cryptocurrency (and specific cryptos like Bitcoin and ETH) and blockchain and easily find hundreds - if not thousands - of recent articles. Those articles run the full spectrum, with some touting cryptocurrency and blockchain as the solutions to most all our problems, and still others telling you cryptocurrency and blockchain are "fads" and will never survive.

\\\ /// CRYPTOCURRENCY, FIAT MONEY, AND LEGAL TENDER

Before we dive into the details, let's talk briefly about money. You may be thinking you don't need an introduction to money, but let's make sure you're comfortable with a couple of terms.

Fiat Money

The first is fiat. ***Fiat money*** is a currency that has been established as money by a government and through government regulation. The government controls the supply of fiat money, establishes monetary policy, sets interest rates, and so on. That's the role of the Federal Reserve in the U.S., and, of course, our fiat money is the U.S. dollar. In Japan, the equivalent of the Federal Reserve is the Bank of Japan and the fiat money is the Yen. For the European Union it is the European Central Bank with the fiat money being the Euro. You get the idea. (The broad general term of *central bank* is used to refer to a specific country's entity that controls its fiat money. The U.S. Federal Reserve, the Bank of Japan, and the European Central Bank are all examples of a *central bank*.)

Legal Tender

Next is ***legal tender***. Legal tender is a form of money that a particular government's courts of law require to be accepted to settle debt. Again, in the U.S. the legal tender is the dollar. (It's entirely possible for a government to create a fiat money that isn't legal tender, but that doesn't make much sense.) Interesting point here. There is no U.S. federal law that requires you or your business to accept folding cash and coins. The *form* of the legal tender that you choose to accept is entirely up to you. Those forms include personal checks, money orders, cashier's checks, wire transfers, credit cards, debit cards, and so on. You can also choose to accept alternative forms of payment including crypto like Bitcoin and ETH, foreign currency, and even fresh eggs and produce.

Cryptocurrency

Okay, so let's now talk briefly about cryptocurrency. A ***cryptocurrency (crypto)*** is a currency in digital or electronic form, with no physical equivalent. Let's use Bitcoin for discussion purposes.

- That cool Bitcoin insignia you see everywhere is something completely made up.
- The Bitcoin insignia is neither copyrighted nor trademarked. It falls under the Creative Commons license, meaning that it can be freely used for commercial and personal purposes.
- There isn't a pile of Bitcoin lying around in a vault somewhere.
- Whenever new Bitcoin is released, there isn't a mint somewhere that produces the equivalent amount in physical form.
- You can't take your digital version of Bitcoin and go down to a bank somewhere and demand a physical Bitcoin. (Actually, you can, but expect to get laughed at.)

So, the real question is whether cryptocurrency is fiat money, legal tender, both, or neither.

And the answers are… it depends, it depends, it depends, and finally it depends.

Sorry, the real answer lies within a CBDC, or central bank digital currency.

Central Bank Digital Currency (CBDC)

A ***central bank digital currency (CBDC)*** is the digital form of a country's fiat money, which is regulated by its (i.e., the country's) central bank. So, will existing cryptos like Bitcoin and ETH ever become the fiat money for a country? The answer is no, because no country's central bank will ever have control over Bitcoin, ETH, or any other existing crypto. No country's central bank can control the supply of existing cryptos like Bitcoin and ETH.

Will existing cryptos like Bitcoin and ETH ever become legal tender for a country? The answer to that question is yes, but it depends on the country. In 2021, for example, the country of El Salvador mandated Bitcoin be accepted as legal tender. [6] The Central African Republic has as well. In those two countries, citizens can settle debt with Bitcoin and the entity issuing the debt is required to accept Bitcoin. It will be interesting to see how legal tender Bitcoin plays out in those countries.

Quick note here… most countries have established laws and regulations regarding the use of existing cryptocurrencies. Countries in which certain cryptos are acceptable (but not required) for use include the U.S., those in the European Union, Australia, Canada, and many others. Some countries, though, have banned the use of existing cryptocurrencies within their borders. A few of those countries include China, Egypt, Iraq, and Morocco.

So, for a country to have "crypto" as its fiat money, it must create its very own central bank digital currency, or CBDC. (It also goes by the term *cryptofiat*, although CBDC is now more pervasive.) In this way, a country's central bank can control the supply of its CBDC, establish monetary policy, set interest rates, etc.

CBDC and its physical equivalent will be the same, holding the same value. $1 in U.S. CBDC will be exactly equivalent to a $1 bill, 4 quarters, or 100 pennies.

Many countries are moving forward to build and issue their own CBDCs. Among them are the U.S. obviously (President Biden signed an Executive Order in 2022 directing the U.S. government to assess the technological infrastructure and capacity needs for a potential U.S. CBDC), Brazil, India, the U.K., China, Bahamas, Jamaica, and Nigeria. The latter four have already released beta versions of their CBDCs, with China being the real leader.

China's CBDC goes by many names… the digital yuan, e-yuan, e-CNY, digital renminbi, and digital RMB. As of January 2022, it was reported that 261 million people in China had set up a digital yuan wallet. That's about 19% of China's total population. [7] However, many of these people probably created a digital yuan wallet in the hopes of winning free cash, with well over the equivalent of $50 million U.S. being given away in various promotions and lotteries.

A CBDC has many advantages. Firstly, in the long run, it's much cheaper to produce and distribute a digital form of cash than it is to produce and distribute a physical form of cash. Consider in the U.S. that it costs a little over 2 cents to mint a penny and a little over 7 cents to mint a nickel. [8] (Hmmm) Most importantly, it combines the safety, convenience, and security of using digital currency with the government's ability to control supply, affect monetary policy, and so on. As we'll discuss later, it's almost impossible to duplicate or counterfeit digital currency.

The supposed promise of cryptocurrency is the technology on which cryptocurrency is based, that is, blockchain and the concept of a distributed ledger.

\\\ DISTRIBUTED LEDGER /// TECHNOLOGY

In Figure 2.1, you can see the relationships between cryptocurrency, blockchain, and distributed ledger technology.

- All blockchain implementations are based on distributed ledger technology.
- But not all implementations of distributed ledger technology are blockchain.
- All cryptocurrencies are based on blockchain.
- But not all implementations of blockchain are cryptocurrency.

Figure 2.1 **Distributed Ledger Technology, Blockchain, and Cryptocurrency**

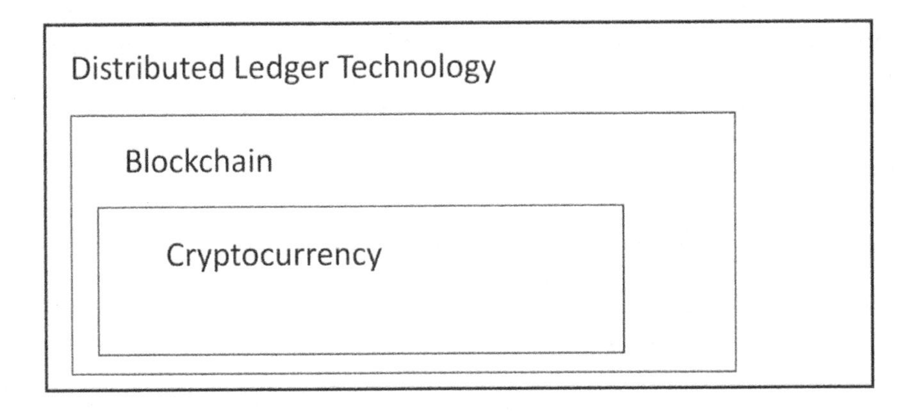

The vast majority of our technology systems today are based on a *centralized ledger technology*, with one "master" set of records stored in a centralized database, or ledger, and one central authority/organization that oversees the quality, updating, and use of the centralized database/ledger.

Your school has a centralized ledger for your classes and grades. The software your school runs provides all the rules for updating and use of that information. SeatGeek likewise has a centralized ledger of concerts, events, tickets, customers, and so on with accompanying software that controls the updating and use of that information. The U.S. banking system is based on a central ledger concept. That's why, when you deposit a personal check from a relative, it takes 2 to 5 days for the check to clear and for you to have access to the money. That check and the movement of money have to be processed centrally. Your bank has no way of knowing if the checking account of your relative (at another bank) has sufficient funds to cover the check.

The IRS, TikTok, Amazon… you get the idea. For almost every environment, there is a "master" set of information and a single authority with complete control over that information.

On the other hand, a ***distributed ledger technology (DLT)*** is an environment in which all nodes on a distributed network have a copy of the master information and must aggregately agree on and approve the use and updating of the information. Thus, each node has a complete and up-to-date copy of the master information and the software for managing the information. As such, the master information (i.e., the ledger) is distributed among everyone. This eliminates the need for a single authority to oversee the information and its use, updating, etc. When the nodes agree to the authenticity of a new transaction, each node updates their respective version of the master information, thus ensuring that every node has an accurate and up-to-date copy of the master information.

\\\ BLOCKCHAIN

Blockchain is an implementation of distributed ledger technology in which information is stored in the form of blocks, with each block holding multiple records or transactions, and with the blocks linked or *chained* together via a unique key or hash (See Figure 2.2). Again, all ***blockchain nodes*** (or simply ***nodes***, the technical term for players and participants who verify transactions, maintain the integrity of the network, and keep an up-to-date copy of the ledger) must agree on the updates to and use of blockchain-stored information. And all nodes have an up-to-date and accurate copy of the blockchain-stored information.

Figure 2.2 **Blockchain**

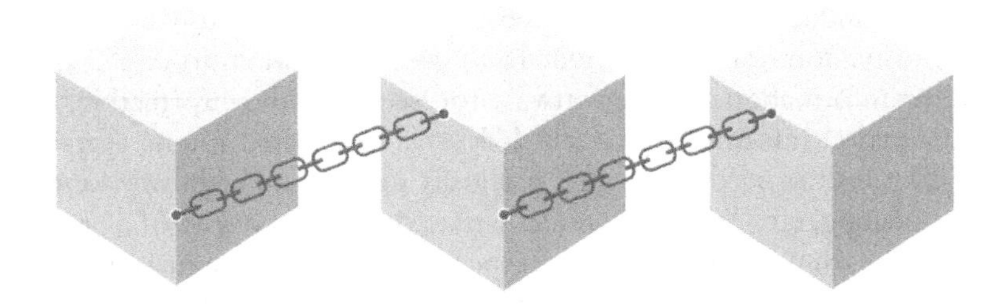

Let's consider a simple example. You and your friends have decided to start your own country, but you don't want to use a centralized banking authority like that of the U.S. Federal Reserve. Instead, you agree that everyone (i.e., the citizens of your new country) will have to approve each financial transaction and that everyone will keep a copy of all the financial transactions. So, all the citizens are the nodes, responsible for verifying transactions and keeping an up-to-date copy of the ledger, and the ledger is *distributed.*

You decide to pay a friend for a product she makes. So, you tell all the nodes/people/citizens that you want to send her $10. Everyone gets word of the proposed transaction, validates that she and you are citizens, and that you have $10 in your account to send. All nodes approve the transaction and then the transaction gets executed, with each citizen updating their copy of the financial database/ledger. Now, you have $10 less and she has $10 more in your respective accounts.

Sounds easy enough in concept, but the technical implementation of a blockchain isn't.

How Blockchain Works

Blockchain is an implementation of distributed ledger technology in which information is stored in the form of blocks, with the blocks linked or chained together via a unique key or hash. Each node has an up-to-date copy of the blockchain.

For purposes of illustration, let's assume the entire U.S. banking and financial systems are now using blockchain. Every bank, savings and loan, credit union, stock brokerage firm, etc. are on the same blockchain system. Each of these is a node, has an up-to-date and accurate copy of the blockchain, and must approve all transactions.

(Note here: In our previous example of your starting a new country, we talked about each citizen being a node. It was obviously purely hypothetical. If and when our banking and financial systems move to blockchain, individuals like you and your friends won't be nodes. That is, we as individuals won't approve every single transaction. That will be the responsibility of nodes and miners, which may include banks, savings & loans, credit unions, brokerage firms, etc. These institutions don't have to be miners but could be. More on miners a bit later.)

So, you write a personal check to a friend. She takes it to her bank to deposit it into her account. Her bank would send out the transaction information to all the other nodes (banks, S&Ls, miners, etc.) who would determine if the transaction is valid using their copies of the distributed ledger:

- Does your bank exist?
- Do you have an account with that account number with your bank?
- Do you have enough money in your account to cover the check?
- Does your friend's bank exist?
- Does your friend have a valid account with her bank?

If all the nodes run through the checklist above (by searching back through the blockchain to, for example, determine if you have opened an account with your bank and have the right account number) and give it a "thumbs up," the transaction gets executed and all the nodes write the detail of the transaction into their distributed ledger copy of the blockchain.

This continues to happen for all financial transactions, until a block gets full. Blocks do have limited space, determined mostly by the size of storage required for each transaction. Once a block is full, a unique hash or key is built that uniquely describes the contents of the block. It is added to the end of the block. That block is then written permanently by all nodes to their respective copy of the blockchain. And then every node starts a new block, with the beginning entry into the new block being the hash or key that ended the previous block. That block-ending hash or key becoming the block-starting entry for the next block is what *chains* the blocks together.

And please do understand that the process we described above is completely handled by software, no human intervention other than the teller at your friend's bank scanning in your check. And the process would be extremely fast, perhaps validating your check and moving the money into your friend's account in a matter of a few hundredths of a second, probably much faster.

Block Hashes or Keys

In blockchain, a ***block hash*** or ***key*** is a unique 256-bit 64-character series that uniquely describes the contents of the block. These work on a simple trap-door principle… you can work forward starting the block content to derive a unique hash or key; but it's impossible to reverse engineer and start with the hash or key and work backward to determine the block content that was used to derive the hash or key. This is a key security concept in blockchain. If someone steals the hash or key for a block, they absolutely <u>cannot</u> use it to determine the contents of the block.

As an example, let's assume we have completed block that has only 2 transactions:

- Arielle sent $100 to Francesca
- Francesca sent $100 to William

Using those 2 transactions, the derived 256-bit 64-character hash or key would be:

f969f3239f5e7e07860cce91b3e32b3c479419dbe4cf5d1ebe0c9e445d44b169

Again, this would be the ending hash for that block and the starting entry for the next block, creating the chain.

Now, let's suppose that a hacker gets into the block and changes the second transaction to reflect that Francesca sent $1000, instead of $100, to William. The new 256-bit 64-character series would be:

7136fc8454c7b0f60acf52a6689b70f930fc665a30b31567a229bd6aced7e620

Those 2 hashes are remarkably different. This is another important security concept associated with using blockchain. If a hacker can ever get into a blockchain and change a transaction, the hash or key changes and no longer matches the starting hash for the next block, thus breaking the chain. This causes a system error.

Recall also that every node on the network has an up-to-date copy of the blockchain. So, if you were a node and got hacked, your copy of the blockchain would not match the copy of the blockchain held by other nodes, creating another system error. Again, this is an important security feature of blockchain. All distributed copies of a blockchain on a network must be identical.

(If you'd like to play around with a 256-bit 64-character key, visit https://passwordsgenerator.net/sha256-hash-generator/.)

Non-Cryptocurrency Implementations of Blockchain

Our discussions of blockchain to this point have focused on cryptocurrency. And blockchain is an essential technology for cryptocurrency; we can't have cryptocurrency without blockchain.

But not all blockchain implementations are cryptocurrency. Indeed, blockchain is applicable to just about any and every use of technology to store, manage, and verify the integrity of transactions and the stored information.

- Voting - The USPS has filed for a patent using blockchain to create, store, and verify voter registration and voting records. [9] Within this system, each citizen will have a unique voter ID and their votes will be stored on the blockchain, with election officials participating in the verification of the votes.
- Diamond Tracking - In 2020, the DeBeers Group announced that it successfully piloted a blockchain-based system for tracking diamonds. [10] The system tracked each diamond from the initial mining of the diamond all the way through retail sale.
- Supply Chain Management - Walmart is using blockchain to manage its supply chain management process. [11] Using blockchain, Walmart has created a system to automate its invoices from and payments to its 70 third-party freight providers in Canada.
- Real Estate - The full gambit of the real estate industry - residential, commercial, vacation rental homes, and even parking spots - is poised to adopt blockchain technology. [12] On the residential front, for example, blockchain can be used to track home ownership, mortgages, lawsuits for payments for work done but not paid, and much more.
- Healthcare - The healthcare industry is similarly exploring the use of blockchain for a variety of applications. [13] Everything from patient medical records, insurance payments, and the tracking of disease outbreaks can be stored and managed on a blockchain.

- Financial Services - Many of our illustrations of blockchain have been around financial services, and there are many more. [14] Some of those include traditional banking services like checking accounts, stock market transactions, identity verification, and credit reporting.

Blockchain may hold great promise for making basic financial services available to the almost 1.7 billion adults worldwide who are ***un-banked***, meaning that they do not have an account with a financial institution or a mobile money provider. [15] Many of these un-banked people cannot afford a traditional account because of the fees. ***Decentralized finance (DeFi)*** is a type of emerging financial technology based on distributed ledger technologies (i.e., blockchain) that may very well eliminate the fees associated with traditional financial account services like checking and savings. DeFi falls within the broader category of ***fintech***, the integration of technologies - like blockchain - that improves the delivery, efficiency, and cost of basic financial services.

The list of promising applications of blockchain goes on and on. Pick an industry, pick any information- or transaction-intensive environment… they can all benefit from blockchain.

Smart Contracts and Dapps

A ***smart contract*** is a piece of software that runs on a blockchain and automates processes, tasks, and the movement of money as certain conditions are met. Within a smart contract, there are a number of nodes that must verify that those conditions have been met.

Suppose, for example, that you want to build a home. You would place money in an escrow account, which the smart contract would manage on a blockchain. As your general contractor and subcontractors complete certain aspects of your home construction, the smart contract would require the appropriate nodes verify the work. Once all nodes verify the work, payment is automatically deducted from your escrow account and sent to the appropriate contractor.

As an illustration, you would have a subcontractor responsible for the electrical work on your home. After that subcontractor completes the work, it would have to be verified and approved by the general contractor, the county inspector, and you as the owner. Once those three nodes approve the work, the electrical contractor is immediately paid by the smart contract using your escrow monies.

Rinse and repeat for plumbing, framing, roofing, carpeting and flooring, painting…

Decentralized apps (Dapps) are also software applications that run on a blockchain. There are some subtle and important differences between smart contracts and Dapps (and a lot of overlap), but we'll not delve into them here.

Dapps include the likes of: [16]

- LBank - a crypto asset trading platform
- PokerKing - online gambling
- CryptoKitties - trading and gaming platform
- Brave - Web browser
- EOS Dynasty - role-playing, player-to-player blockchain-based game
- TRACEDonate - for tracking how your money gets used when you make a charitable donation
- Circulor and Chainyard - supply chain management
- Aeve - for lending and borrowing cryptocurrency assets (DeFi)
- Compound - for earning interest while borrowing and lending cryptocurrency assets (DeFi)
- Axie Infinity - combines elements of Pokemon and crypto collectibles

Decentralized Autonomous Organizations (DAOs)

According to Cooper Turley, a ***decentralized autonomous organization (DAO)*** is an Internet community with a shared bank account. [17] A DAO is basically a group of people who decide to pool their money for any number of purposes and establish a set of rules by which the group will operate.

Creating a DAO is rather like forming a partnership with several of your friends. You pool your money into a single account, and then you establish rules by which your partnership will operate. What makes DAOs particularly appealing is that you automate your operating rules in the form of a smart contract on a blockchain.

Let's suppose your partnership wants to get into investing in rental real estate. You could establish automated rules around the approval of a purchase, for example that 100% of the partners must approve the purchase of a new piece of real estate. You could establish rules around the distribution of profits, for example that 20% of the profits will be held in escrow in case you need money for emergency repairs and maintenance. You could establish rules around the sale of a piece of property, for example that at least 90% of the partners must approve the sale.

Again, all these rules would be automated on the blockchain. Every partner would receive notification of the pending execution of a rule and be provided the right to vote on its approval.

Obviously, DAOs are mainly prevalent in the blockchain and cryptocurrency worlds. A DAO needs software that enables the automation of rules, that is, smart contracts that run on the blockchain. And, as most DAOs have some sort of financial mission (investment, etc.), the money medium is most appropriately cryptocurrency because of its each of use on blockchains and in smart contracts.

Some examples of well-known DAOs include: [18, 19]

- PleasrDAO - invests in NFTs (discussed later) and other digital investments
- HerStory - funds projects by black women and non-binary artists
- Komorebi Collective - funds women and non-binary crypto founders
- Friends with Benefits - pay-to-enter exclusive social club
- MetCartel Venture - invests in early-stage decentralized applications
- Decentraland - 3D virtual world browser-based game
- Illuvium - open-world fantasy battle game

\\\ CRYPTOCURRENCY

Okay, we've been talking a lot about cryptocurrency to illustrate distributed ledger technology and blockchain. Let's now take a deeper dive into that fascinating topic.

Public and Private Keys

As we've already stated, when you get crypto (a friend sends it to you, you buy some on an exchange, etc.), you don't get - and you can't get - a physical coin. All cryptocurrency are completely digital. As well, all cryptocurrency use cryptography to secure, validate, and authenticate transactions. That's why you see "crypto" in front of the word currency.

Formally, the type of cryptography that all cryptos use is called ***Public Key Cryptography (PKC)*** or ***Asymmetric Encryption***, a trap-door approach that makes use of public keys and private keys to secure transactions.

A ***public key*** is a key on a blockchain that is publicly known and is used mainly for identification. In the crypto world, someone has to know your public key in order to send you cryptocurrency. A ***private key*** is something very private and is an electronic key that gives you the sole and exclusive right to use your cryptocurrency. Think of a public key as something on a blockchain everyone can see and thus identifies that you are the owner of cryptocurrency. Your private key is what gives you the right to use that cryptocurrency.

So, when you buy cryptocurrency (or a friend sends it to you), you receive a private key for it, giving you the right to use it, send it to someone else, trade it for another cryptocurrency, etc.

To keep your cryptocurrency safe, you need to securely store your private keys in a cryptocurrency wallet or on a cryptocurrency exchange.

Cryptocurrency Wallets and Exchanges

There is definitely a difference between a cryptocurrency exchange and a cryptocurrency wallet. A ***cryptocurrency exchange*** is a marketplace on which you can buy and sell crypto. Popular cryptocurrency exchanges include Crypto, Coinbase, Exodus, Binance, Kraken, Robinhood, eToro, and Gemini. Think of a cryptocurrency exchange as your bank holding your bank accounts, and thus your money. All cryptocurrency exchanges offer wallet capabilities.

A ***cryptocurrency wallet*** is a software system that allows you to store and manage your cryptocurrency. Again, all exchanges offer wallet capabilities. The downside to most of these is that the exchange holds and manages your private keys to the cryptocurrency you own. Exchanges can be hacked and your private keys can be stolen. If so, you may no longer own that cryptocurrency because the hacker can use your stolen private keys to move the crypto elsewhere. (It's gone.)

In 2021, Liquid, a Japan-based cryptocurrency exchange, was hacked with the hackers gaining access to and making off with almost $100 million in cryptocurrency. [20] The largest hack/theft to date was against the Ronin Network in March 2022. [21] In that heist, hackers made off with $625 million worth of cryptocurrency. The second largest known heist of cryptocurrency ends with an interesting twist. In August 2021, a hacker made off with a little over $600 million from the Poly Network. The Poly Network made an appeal on social media for the return of the crypto and set up several addresses to which the crypto could be returned. After only a few days, the hacker had returned over $300 million in crypto, along with an explanation that they were only doing the hacking for "fun."

So, many experts recommend that you store your cryptocurrency in a cold cryptocurrency wallet, as opposed to a hot cryptocurrency wallet. A ***hot cryptocurrency wallet*** (also ***hot wallet*** or ***hot storage***) is an "always-on" cryptocurrency wallet connected to the Internet, thus enabling real-time and live transactions. So, if you use Coinbase as your cryptocurrency exchange and use its Internet-based cryptocurrency wallet for storing your cryptocurrency, you are using a hot cryptocurrency wallet.

On the other hand, a ***cold cryptocurrency wallet*** (also ***cold wallet***, ***cold storage*** or ***hardware-based cryptocurrency wallet***) is a type of offline cryptocurrency wallet that you must explicitly "connect" to the Internet in order to use the cryptocurrency stored on it. The most popular type of cold cryptocurrency wallet is a USB-based hardware wallet. That is, all the wallet software and the private keys for your cryptocurrency are stored on a USB drive. Disconnect it from your computer and it can no longer be accessed by the Internet. To use it and the cryptocurrency, you must plug the flash drive into your computer.

Popular hardware-based cryptocurrency wallets include Ledger, Trezor, KeepKey, Ellipal, Secalot, BitBox02, D'CENT Biometric Wallet, and Keystone. All these allow you to move your cryptocurrency (i.e., private keys) offline. When you move your cryptocurrency there and then unplug it, no one can access your private keys to your cryptocurrency. Ever.

Of course, if you happened to lose that flash drive, it's the same as losing your physical-money wallet or your money clip. Your money is gone. Bye-bye.

Stablecoins

Because cryptocurrencies like Bitcoin and ETH are not regulated by a particular government through monetary policy, they tend to be extremely volatile. Take a look at the table in Figure 2.3. The range for each during the year was substantial. Bitcoin ranged from a low of $16.5k to a high of $44k. Likewise, ETH ranged from a low of $1,195 to a high of $2,380.

Figure 2.3 **Highs and Lows of Bitcoin and ETH in 2023**

Value	Bitcoin	ETH
January 1, 2023	$16,600	$1,199
2023 Low	$16,500	$1,195
2023 High	$44,000	$2,380
December 31, 2023	$42,000	$2,299

Because of this volatility and rather extreme change in value, it's hard to imagine that any government would adopt either of the two most popular cryptocurrencies as fiat money. (As well, since a government doesn't control the supply of either cryptocurrency, it cannot regulate them through policy.) To reduce the volatility of cryptocurrency and make it act more like traditional fiat money, people have invented stablecoins.

A ***stablecoin*** is a type of cryptocurrency that is pegged or tied to an external asset like the U.S. dollar or gold. This helps stabilize the price of the stablecoin. Popular stablecoins include (all tied to the U.S. dollar): [22]

- Tether (UDST)
- Dai (DAI)
- Binance USD (BUSD)
- TrueUSD (TUSD)
- USD Coin (USDC)
- Gemini Dollar (GUSD)
- BitUSD (BitUSD)
- USD Digital (USDD)

You can buy any of these stablecoins, and many others, on popular cryptocurrency exchanges.

Of course, the biggest advantage of owning stablecoin is that $1 in Tether, for example, will still be worth approximately $1 in the future. The purchasing power of that $1 may have declined because of inflation. But the same is true for your traditional money in your bank account.

Cryptocurrency Mining

During the creation of a block, the nodes on the blockchain are responsible for validating transactions, creating the ending hash for a block, keeping an up-to-date copy of the blockchain, etc. Some of these nodes are referred to as miners. ***Cryptocurrency mining*** (***mining***) then is the process of verifying transactions in a block, creating the ending hash for a block, and maintaining the integrity of the network. The question then becomes, "How do miners get paid for all that work?"

There are two approaches to paying miners for their work, either proof of work or proof of stake. Proof of work is the older of the two.

In ***proof-of-work mining***, miners literally race to see who can be the first to create a verifiable unique hash that describes all the information in a block. So, when a block fills, the race is on among the miners. This involves solving a hugely complex mathematical problem using the information in a block to create a unique hash that uniquely describes the information in a block, such that even the slightest change to the block of information will cause the hash to be different.

When a miner is the first to solve that math problem, they send it out the hash to all other miners for verification. The other miners verify the hash, and then all miners/nodes write the block to their respective distributed copy of the blockchain.

The miner that solves the problem and builds that hash is rewarded monetarily. Using Bitcoin as an example, the miner who "wins the race" by solving the computational problem first is paid 6.25 bitcoin (as of late 2023). And, this is completely new bitcoin. That's how new bitcoin enters the market; new bitcoin is generated and entered into the system/market by paying the miner who wins the race.

In ***proof-of-stake mining***, miners "stake" a certain amount of cryptocurrency to guarantee their work in creating the unique hash for a block. The system randomly chooses several of these staked miners. Once a certain number of the chosen miners verifies the block (including the ending hash), all nodes/miners write the block to their respective distributed copy of the blockchain.

(If a chosen miner provides incorrect verification of a block - that is, the wrong ending hash - that miner relinquishes their staked cryptocurrency.)

The miners who were chosen and verified the block are then paid with completely new cryptocurrency, just as in the case of proof-of-work mining.

Proof-of-stake is the chosen mining method for many cryptocurrencies including ETH, with each miner required to stake 32 ETH cryptocurrency coins.

Proof-of-Work Mining and Bitcoin
Bitcoin uses the proof-of-work mining concept. When Satoshi Nakamoto wrote the original paper on Bitcoin, he/she/they outlined the payment mechanism for mining based on the proof-of-work model. Satoshi also included the fact that there would only ever be 21 million Bitcoin.

During the early years of Bitcoin, miners were paid at a rate of 50 bitcoin per block. That has been halved roughly every 4 years. As of 2023, Bitcoin miners are paid at a rate of 6.25 bitcoin per block. 6.25 bitcoin may seem like a lot of money, as it was approximately $250,000 in late 2023. But (and it is a big but), the initial technology investment into all the hardware and software for mining Bitcoin will easily be in the millions of dollars. (Seriously.) It takes a lot of computing horsepower and electricity to be a Bitcoin miner. And there's no guarantee that you'll win the race. There are probably upwards of a million Bitcoin miners today worldwide all racing to be the first to verify a block and build the unique hash. If you win, you get paid approximately $250,000. If you lose, you just spent a lot of money without getting any return.

The Inventors of Cryptocurrency

As of early 2024, there were over 23,000 different cryptocurrencies in existence. [www.coinmarketcap.com]

Because cryptocurrency is not fiat money (established and legislated by a government), quite literally any person can invent their own cryptocurrency. That may sound appealing but the technical complexities of doing so are enormous. And getting market acceptance is even more difficult.

So, who are the inventors of some of the more popular cryptos out today? They include:

- Bitcoin (BTC) - Satoshi Nakamoto (more on that fascinating story in a moment)
- Ether (ETH) - Vitalik Buterin
- Stellar (XLM) - Jed McCaleb
- Binance Coin (BNB) - Changpeng Zhao
- Cardano (ADA) - Charles Hoskinson
- Polkadot (DOT) - Gavin Wood
- Chainlink (LINK) - Sergey Nazarov
- LitCoin (LTC) - Charlie Lee
- Bitcoin Cash (BCH) - Craig Wright
- Tether (USDT) - Brock Pierce, Craig Sellars, and Reeve Collins

Satoshi Nakamoto gets the nod for really launching the entire cryptocurrency craze. Satoshi wrote a white paper in 2008 that described Bitcoin as a peer-to-peer (i.e., no centralized authority or server) electronic cash system. Interestingly enough, no one knows who Satoshi Nakamoto is. No kidding. We don't know nationality, gender identification, or if it's a single person or a group of people. (Craig Wright, the creator of Bitcoin Cash, claimed to be Satoshi in 2016 but later backed away from that claim.)

Getting Started Investing in Cryptocurrency

It's pretty simple. We would recommend that you register for an account with a cryptocurrency exchange like Coinbase. Once you register (it's free), you'll have a web-based hot wallet and you can start to invest. Some things to keep in mind:

- Most exchanges require you to be at least 18 years of age.
- All exchanges have country limitations; some countries won't allow you to buy cryptocurrency.

- When you do sell, all exchanges want tax information; you will get a 1099 (in the U.S.) because the U.S. government currently treats cryptocurrency as an investment.
- Look at the fee structures carefully. These differ from exchange to exchange.

Most importantly, and we're very serious about this… *don't bet hurt money.* Treat your investment in cryptocurrency as purely speculative. If you're on a tight budget right now, cryptocurrency is a dangerous play because of its volatility. But, if you have a hundred bucks lying around, then find a few cryptos that interest you and spread your money among them… something about, "Don't put all your eggs in one basket." And, be prepared to just leave it alone. See how you're doing 6 months from now.

The Internet's Influence on Cryptocurrency: The Rise of Dogecoin

While the topic of crypto is complex, its functions are not all too different from a stock investment. The value of each respective holding is influenced by the investor's future outlook and the investment's popularity. You may have witnessed the rise of GameStop and how the sudden popularity of the stock on the internet actually saved the company from going under. While it serves as a meme, it's a great example of how internet popularity can influence a stock's value.

In the crypto world, this was Dogecoin. Again, the idea of investing in Dogecoin started as a meme. It was an extremely cheap crypto and had a cool looking logo. Yet because of its internet popularity, people like Elon Musk took notice and provided endorsement. Dogecoin today is one of the most successful cryptos, up 4,500% from 5 years ago. At its peak, it was up almost 15,000%. (Keep in mind this was only about $0.64 a coin at this peak.) While this is not viable for all cryptos and stocks, internet popularity and "celebrity endorsement" do play a role in which of these investments can skyrocket. Keep this in mind as you see different trending cryptos on the rise.

The Benefits of Blockchain-Based Cryptocurrency

So, now you have some idea of how blockchain-based cryptocurrency works. And blockchain is essential for cryptocurrency, and its many other applications. For purposes of our discussion of the benefits and advantages of blockchain-based cryptocurrency, let's shift our example to Ether, the second most popular crypto. Ether is often just referred to as ETH (long e-th).

No Double Spending: Let's assume you buy a single ETH coin. Every ETH coin (and fractional part of an ETH coin), has a unique ID, of sorts, that is, the private key. (Think of this unique ID as being analogous to a serial number on a dollar bill. Each is unique.) Let's say the unique ID for your ETH coin is 1234. When you buy ETH 1234, your ownership is stored on every distributed copy of the blockchain on the network. If someone else were to try to spend your ETH coin (1234) or claim ownership to it, the nodes on the network would invalidate the transaction because you are the owner and not the other person.

So, no one can make a copy of your ETH coin. There can never be 2 ETH coins with the unique ID of 1234. That has significant, positive ramifications with respect to eliminating counterfeiting.

Immutability: Immutability is a characteristic of a blockchain that basically says once a transaction has been approved and recorded it cannot be changed or altered in any way. So, there are no erasers in the blockchain world. If, by accident, you enter an incorrect transaction, such as selling 2 widgets when you actually sold 20, you would have to submit another transaction to reflect the sale of the other 18. You cannot "erase" the 2 in the previously approved transaction and write in the 20. If you were to do this, the hash for that block would change and no longer match the beginning hash in the next block, resulting in a system error.

Non-Fungible Tokens (NFTs)

A ***non-fungible token (NFT)*** is a blockchain-based token that is non-interchangeable and represents ownership. (That's probably about as clear as mud.) Think of it in economic or financial terms. Traditional currency is fungible. Think about a $20 bill. It represents that you have $20 in value for purchasing something. And you can exchange a $20 bill for 2 $10 bills. That exchange ratio has been formally defined by the U.S. government and includes all denominations of coins and paper money.

Non-fungible tokens, while they can be extremely valuable, are very different. Someday when you buy a car, you may get an NFT representing ownership of that car, instead of the traditional car title. The NFT will be stored on a blockchain and represents that you own that car. When you sell the car, you'll need to know the new owner's public key and you'll use that key to send the new owner the NFT for the car and the private key that gives them the exclusive right to the car.

Likewise, there is no formal exchange ratio for NFTs. There's nothing out there that says your 2021 Toyota Corolla is worth exactly 2 2011 Honda Civics. If you wanted to "swap" your 2021 Toyota Corolla for 2 2011 Honda

Civics, you would have to sell your car (and the representing token or NFT) to someone and in turn they would sell you the 2 2011 Honda Civics (and the representing tokens or NFTs.)

What really became popular in 2021 was the use of NFTs to represent ownership of digital content like digital art, video clips, memes, virtual real estate in metaverse, trading cards, and the like. When we wrote this book, the most expensive NFTs ever sold for digital content included: [23]

1. Murat Pak's *The Merge* - $91.8 million
2. Beeple's *Everydays: The First 5000 Days* - $69.3 million
3. Pak and Julian Assange's *Clock* - $52.8 million
4. Beeple's *HUMAN ONE* - $28.95 million
5. CryptoPunk *#5822* - $23.7 million
6. CryptoPunk *#7523* - $11.75 million
7. TPunk *#3442* - $10.5 million
8. CryptoPunk *#4156* - $10.26 million
9. CryptoPunk *#5577* - $7.7 million
10. CryptoPunk *#3100* - $7.58 million

Those numbers seem almost unbelievable, but they are true. All of those dwarf the high-profile sale of a video clip of a massive dunk by Lebron James. NBA Top Shot sold it for a little over $200,000.

This is truly the wild, wild west. In September 2021, Snoop Dogg announced he was selling 1,000 NFT passes for people who wanted to virtually party with him in Sandbox, a popular metaverse destination. [24] That same year, an investor paid $1.23 million to own the virtual land (via an NFT) next to Snoop Dogg in Sandbox. That's a lot of money to be Snoop Dogg's <u>virtual</u> next-door neighbor. [25]

You, too, can mint your own NFTs for just about any digital content you care to create. If you do, you'll want to sell them on an NFT exchange. The more popular ones include: [26]

- OpenSea
- Axie Marketplace
- Larva Labs/CryptoPunks
- NBA Top Shot Marketplace (NBA-specific NFTs only)
- Rarible
- SuperRare
- Foundation
- Nifty Gateway
- Mintable
- Theta Drop

Many of those NFT exchanges have tools so you can build your own digital content and NFTs.

\\\ THE END GAME:
/// WHAT THE FUTURE MIGHT HOLD

Cryptofiat

In the next 3 to 5 years, a major country will announce that its fiat money is now a cryptocurrency. We believe that first big country will be China. For the past several years, China has been piloting its own national cryptocurrency.

It's just a matter of time - and not much time - before China adopts the digital yuan as its fiat money.

Other countries will follow.

Information Storage and Updating Will Evolve to Blockchain

Way back in the 1970s and 1980s, information storage evolved from *files* to *databases.* In fact, the term database is ubiquitous in information storage and simply refers to a collection of files. So, files didn't go away; they simply morphed into databases.

The same will be true for databases and blockchain. Databases aren't going away. But they will morph into blockchain-based storages of information.

Recommendation: If you want to be in the IT field, take a course in blockchain. It's the future.

PayPal, Venmo, and Others Will Pivot to Cryptocurrency

Services like PayPal and Venmo are peer-to-peer financial networks. They enable you to easily move money - the U.S. dollar for example - to other people.

That *money* will become cryptocurrency. (PayPal already lets you buy and sell cryptocurrency.)

The Death of Folding Cash and Coins

This certainly seems inevitable. It costs a lot of money to mint folding cash and coins. And, when fiat money becomes cryptocurrency, folding cash and coins will become those seldom-seen oddities. It will happen in your lifetime. Just remember, you heard it here first.

CHAPTER 3

Artificial Intelligence

A Cat Mistaken for Guacamole

The movie became an instant cult classic... an AI (artificial intelligence) got so smart that it understood the concept of procreation, locked a woman inside her home, and impregnated her resulting in her having a computer baby. Sound familiar? Probably not. The movie was *Demon Seed*, and it came out in 1977, some 45+ years ago.

Of course, you've watched untold numbers of more recent movies that depict AI... the *Terminator* series, the *Matrix* series, the *Alien* series, the *Iron Man* series (loved J.A.R.V.I.S.)... the list goes on and on. It seems Hollywood has an unbelievable fascination with artificial intelligence. Of course, Hollywood produced those movies because we as a society are even more fascinated by artificial intelligence.

As an academic discipline and field of study, artificial intelligence has been around since the 1950s. Some 70 years later, a Harvard professor showed how Google's AI-based image recognition technology could be (easily) fooled into believing that it was looking at guacamole, when, in fact, it was an image of a cat. [1] Needless to say, we have a ways to go before artificial intelligence becomes as smart as us humans.

As a formal definition, let's say that ***artificial intelligence (AI)*** is a machine that mimics human intellectual tasks such as learning, problem solving, and social interaction. If you happen to search on artificial intelligence, we can guarantee that you won't find that exact definition. Instead, you'll find hundreds of different definitions. But they all focus on the same thing, making machines seem like they have human intelligence.

Just a word of caution before we dive into artificial intelligence. Without a doubt, AI is the most complicated of all the 4th industrial revolution technologies we'll be discussing in this book. Volumes have been written about the subject. Universities and colleges offer entire degree programs in AI, many at the masters and PhD level. We checked out the Wikipedia page for AI in early 2024, and it included over 600 explanatory notes, references, and further readings.

Big and complicated field.

We're going to try to cover it in less than 25 pages.

\\\THE AI SPECTRUM:
///FROM SINGLE TASK TO SINGULARITY

The spectrum of AI applications, research, and dreams goes from AI performing a single task to AI achieving singularity. Most commonly, AI is presented in terms of the 8 categories in Figure 3.1. But these categories are not mutually exclusive, which makes this field even more confusing and complicated. We'll explain the overlaps after introducing each category to see if we can reduce the confusion and complications. No guarantees, but we'll try.

Figure 3.1 **The Artificial Intelligence Spectrum**

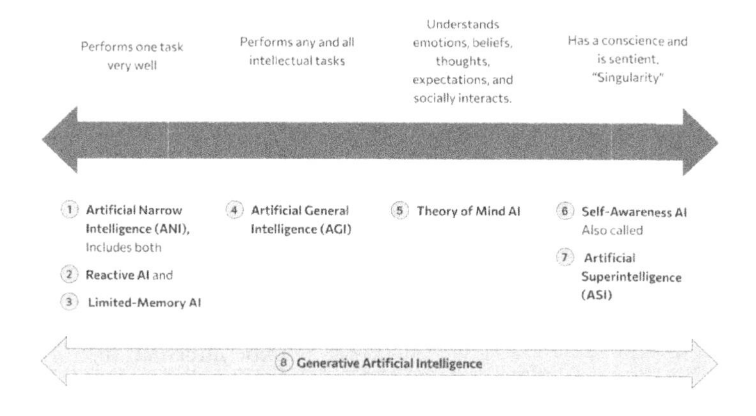

Artificial Narrow Intelligence

On the far left is ***artificial narrow intelligence*** (***ANI*** or ***narrow AI***), an AI that performs one and only one task. Seems rather simple and mundane, but this is where almost all current successful implementations of AI fall, getting a machine to perform one intellectual task very well.

Artificial narrow intelligence includes the likes of Netflix's recommendation engine, car diagnostics software, chatbots, virtual assistants, spam filters, image recognition, handwriting recognition, medical diagnosis, and even autonomous vehicles. Each of these performs a single task. Netflix's recommendation engine, for example, helps you hopefully find a movie that you'd like. It can't tell you which books to read or classes to take in school. It can only recommend what video you might like to watch. Period.

Reactive Artificial Intelligence

Reactive artificial intelligence is a subset of artificial narrow intelligence in which the AI is programmed to provide a predictable outcome based on the inputs it receives.

- The same set of inputs will always produce the same outcome.
- It will always respond to the same situation (inputs) in exactly the same way each and every time.
- It cannot learn to handle new inputs, nor can it adapt itself to new situations.
- It cannot *learn.*

Let's give you an example of a reactive AI for determining what to do when approaching a traffic light at an intersection. You can build a reactive AI for this task in the form of rules as in the table below.

	RULE/QUESTION	YES	NO
1	Is the light red?	Go to Rule #2	Go to Rule #3
2	Do you have time to stop?	Stop	Accelerate through the intersection and hope you don't crash.
3	Is the light yellow?	Go to Rule #4	Proceed through the intersection.
4	Do you have time to stop?	Stop	Accelerate and proceed through the intersection.

Let's work through the example with the light being green. The AI would check the first rule, "Is the light red?" The answer is no, so it proceeds to rule #3. It checks that rule, "Is the light yellow?" The answer is no, so it would tell you to proceed through the intersection.

Similarly, if the light were red, the AI would check the first rule, get a "yes" answer, and then proceed to rule #2. At rule #2, the AI would have to determine - based on your speed and proximity to the intersection - if you have time to stop. If the answer is yes, it would instruct you to stop. If the answer is no, then of course you may have a serious problem.

This type of AI tool is called an expert system. An ***expert system*** uses a series of if-then rules to work through a set of inputs to determine what action to take. (Expert systems have been around quite some time. These types of systems were developed way back in the 1980s using a language called LISP, which stands for **LIS**t **P**rocessor.)

In this simple example, you can see that a reactive AI will always arrive at exactly the same outcome based on 2 simple inputs, the color of the light and whether or not you have time to stop.

You can also see that this AI cannot take into account any other inputs, such as weather, the presence of pedestrians, a left-turn signal, what other vehicles are doing, the presence of emergency vehicles, and so on. It has been programmed to *react* to only 3 colors of a traffic light and whether or not you have time to stop.

If you wanted this AI to take into account the presence of emergency vehicles, for example, you would have to change the way the AI works by altering the if-then rules and adding a couple of more. It would look like the set of rules below.

	RULE/QUESTION	**YES**	**NO**
1	Is an emergency vehicle in the area?	Go to Rule #6	Go to Rule #2
2	Is the light red?	Go to Rule #3	Go to Rule #4
3	Do you have time to stop?	Stop	Accelerate through the intersection and hope you don't crash.
4	Is the light yellow?	Go to Rule #5	Proceed through the intersection.
5	Do you have time to stop?	Stop	Accelerate and proceed through the intersection.
6	Is the emergency vehicle traveling parallel to you?	Pull over at least one lane and wait for the emergency vehicle to pass.	Stop where you are and wait for the emergency vehicle to pass.

Examples of reactive AI include Deep Blue (IBM's AI-based supercomputer that beat Garry Kasparov, the world champion chess player, in 1997), spam filters for our email, car engine diagnostic software, Google's AlphaGo, and

Netflix's recommendation engine (and most all the recommendation engines out there).

Netflix's recommendation engine uses several mathematical and statistical models and rules to make recommendations for you. The more movies you watch, the more reviews you provide, and the more searches you perform on Netflix, the better the system gets at making recommendations for you. That may seem like "learning," but it isn't. It simply has more data at its disposal to make a better recommendation. For example, if you've only watched 2 movies on Netflix, one drama and one western, then Netflix's recommendation engine has a 50/50 chance in offering recommendations of what else you might like to watch. But, if you watched 10 movies, 1 drama and 9 westerns, then Netflix's recommendation engine will simply recommend more westerns. Simple math and probability.

Limited-Memory Artificial Intelligence

Limited-memory artificial intelligence is a subset of artificial narrow intelligence in which the AI learns from historical/past data to make better decisions. So, the more limited-memory AI is used, continually learning and adapting itself, the better it gets at making its decision.

Autonomous vehicles definitely fall into this category of AI. Previous to buying your autonomous vehicle, its AI has been trained in a simulated environment. It's driven perhaps millions of miles on a simulator to learn how to drive, react to different road conditions, handle traffic congestion, and so on. As you drive it (or allow it to drive itself), it continues to learn, getting better at the task of driving.

In this way, limited-memory AI is actually adapting itself to get better at the task of driving. This is analogous to how you drive over time, continually learning and adapting your driving technique to get better.

Limited-memory AI is the dominant category of AI. As it is so dominant, we'll come back to this category of AI in a moment.

Artificial General Intelligence

Artificial general intelligence (***AGI***) is an AI that can take what it knows and has learned for one situation and apply it to a new situation. These AI will function completely like a person, intellectually. They will be able to build competencies in situations without being explicitly programmed for them. They will be able to generalize across domains, adapting their current knowledge to new situations. When using the term "new situation," we're not referring to good weather versus driving in the rain. We're referring to an AI learning to drive a car and then being able to adapt that knowledge to literally teach itself how to drive a motorcycle or speedboat.

This is an important goal in the field of AI, and we are not there yet. Indeed, many people don't believe we'll ever be able to create a general AI that can adapt to new problems and situations, without some sort of pre-programming by us (humans). It remains to be seen.

But researchers and scientists all over the world are working daily to make general artificial intelligence a reality.

You've probably played a lot of different types of card games, including perhaps Go Fish, Gin, Hearts, Spades, Poker (in all its variations), Solitaire (in all its variations), and so on. As you learned these games, you exhibited *general intelligence*. That is, you were able to take your knowledge of a deck of cards - 4 suits, 13 cards in each suit, the hierarchy of cards in a suit, and even some strategies like being void or long in a particular suit - and apply that knowledge to a new card game with a new set of rules. So, you were able to learn the new card game quickly because you already had some experience playing other card games.

It is this type of general intelligence that gives us the ability to quickly learn to perform new intellectual tasks, because we can adapt our current knowledge to a new intellectual task. You learned math this way. You first learned to add. You learned to subtract by understanding that, because 3 + 2 = 5, then 5 - 2 = 3 and 5 - 3 = 2. You also learned multiplication as repetitive addition. That is, 8 * 5 is the same as 8 + 8 + 8 + 8 + 8. Division was easy (relative term) to learn because you already knew subtraction and multiplication.

Artificial general intelligence requires that an AI be able to take its knowledge from one domain and apply and adapt it to another. It requires that an AI adapt to constantly changing environments and completely new and unforeseen circumstances. It requires that an AI foresee a future based on its knowledge of the present and the past.

In short, artificial general intelligence is human intellect, all of it.

Again, long way off.

From Tic-Tac-Toe to Global Thermonuclear Warfare

In the 1983 movie *Wargames*, an AI named Joshua took over the U.S missile systems with the intent of launching a nuclear attack on the Soviet Union. It simulated many scenarios for doing so and could never come up with a great solution. So, it played tic-tac-toe against itself millions of times, and came to understand that the cat always wins, if the players understand the strategies and are equally adept at the game. From that, it extrapolated that there could be no winner in global thermonuclear war.

That's (artificial) general intelligence to the extreme.

Most likely at some point in your life, you learned to play tic-tac-toe in such a way that no one ever won, given, of course, that you were playing against someone else with the same knowledge. Did that lead you to an epiphany that global thermonuclear war was a no-win scenario?

Theory of Mind Artificial Intelligence

Theory of mind artificial intelligence is an AI that interacts socially and can discern and respond appropriately to people's emotions, beliefs, thoughts, expectations, and facial expressions. It takes AI beyond intellectual tasks and into the realm of "understanding" people.

Theory of mind (ToM) has its foundations in psychology. In psychology, theory of mind refers to your capacity to understand other people and assign mental states to them. Conveniently, you can think of theory of mind as having the capacity for sympathy or empathy. It's not a directly analogous and 100% accurate statement, but it should help you understand how theory of mind differs from intellectual tasks such as determining whether a photo is that of a cat or guacamole.

Like artificial general intelligence, we're a long way from theory of mind AI becoming a reality. But there are a couple of early limited successes in this space. Check out:

- Sophia, a humanoid robot designed by Hanson Robotics. In 2017, Sophia became the first robot to receive country citizenship, that of Saudi Arabia. [2]
- Kismet, a robot head designed by Cynthia Breazeal at MIT, that recognizes facial emotional signals of people. [3]

- ToMnet, a theory of mind neural network being developed by Google DeepMind, that will use a small number of behavioral observations to predict a person's characteristics and mental states. [4]

Sophia and Kismet have many, many videos, articles, and the like. Just do a quick search.

Self-Aware Artificial Intelligence

Self-aware artificial intelligence is an AI that is for all practical purposes a person. It will exhibit artificial general intelligence, the ability to adapt its knowledge to new domains. It will exhibit theory of mind, which we just discussed.

A self-aware AI will have its own emotions and be aware of its emotions. It will be sentient. It will have a conscience. It will understand the need to procreate. It will struggle with ethical decisions. It will probably even act irrationally at times, just as we do, and later apologize.

Self-aware AI is purely hypothetical. We can't even build the basic infrastructure of a self-aware AI with our current set of technologies.

Long, long way off if we ever get there. And that's a big IF.

Artificial Superintelligence

Artificial superintelligence (ASI) is the most far-reaching view of the potential of AI. An ASI would be the most intelligent entity on the earth, making better and faster decisions than all humanity combined in all aspects of life. ASI is popularly referred to as *singularity*, the point at which machines become smarter than us.

Long, long way off if we ever get there. And that's another big IF, an even bigger IF than that for self-aware artificial intelligence.

Generative Artificial Intelligence

Generative artificial intelligence (generative AI or ***GenAI)*** is an AI that can dynamically create (new) output, usually based on one or more prompts.

The most well-known generative AI is ChatGPT. In fact, you may have already used ChatGPT to help you write a school assignment. (And, perhaps, your school has banned the use of such tools.) ChatGPT has been trained using millions of pieces of textual data… books, monographs, web content, etc. To use it, you provide a textual prompt of what you want your school assignment to address, important points to cover, length, perhaps style, and so on.

You may also be familiar with the likes of DALL-E, Midjourney, Adobe Firefly, and Stable Diffusion. These are all text-to-image generative AI tools. These have all been trained using millions of images (and their descriptions and/or captions). To use one of them, you provide a textual prompt of what you want your image to look like.

Prompt engineering is the process of creating and structuring a textual prompt that can be interpreted by a generative AI tool. The more complete and informative your prompt the more likely the generative AI tool will be at producing what you want.

Many people confuse generative artificial intelligence with a large language model. A ***large language model (LLM)*** is a subset of generative AI that has been trained using textual data to create (new) text. So, ChatGPT (and others such as Character.ai and QuillBot) is a large language model. On the other hand, generative AI tools that focus on creating images (DALL-E, Midjourney, and others) are text-to-image systems. That is, they take in a textual prompt and produce an image. Strictly speaking, generative AI tools that produce images (and other non-textual output) are not large language models. On the front end, these generative AI tools do use some large language model processing to understand the textual prompt, but their end result is to produce something other than textual output.

The major categories of generative AI tools include:

1. Large language model (text-to-text): ChatGPT, Character.ai (a chatbot that can generate textual responses in a chat environment), QuillBot, BLOOM, and Gemini (the latter three are similar to ChatGPT)
2. Image production (text-to-image): DALL-E, Midjourney, Adobe Firefly, and Stable Diffusion are the major ones.
3. Programming code production (text-to-code): GitHub Copilot, Amazon CodeWhisperer, tabnine, codeium, and specialized versions of ChatGPT.
4. Audio production (text-to-audio): ElevenLabs, Meta Platforms' Voicebox, MusicLM, and MusicGen.
5. Video production (text-to-video): Make-A-Video by Meta Platforms, and Gen-1 and Gen-2 by Runway.

Those are the major categories but many more are emerging in the spaces of poetry, robotics, and even molecules.

If you're wondering how generative AI tools "learn," consider the following sentence: Rosalyn sat quietly on a ____________ and reflected on the day's events. In the case of a large language model, it has learned what word or words most likely follow other words. In the case of the blank in the sentence, the LLM would use something like a park bench or perhaps a lounge chair. It would not, on the other hand, use something like a cactus or a roller coaster. Large language models make these connections among words via a neural network, which we'll discuss in an upcoming section.

A couple of years ago, when we wrote the 2022 version of this book, generative AI didn't even make the list of the categories of artificial intelligence. Today, generative AI is the most talked about tool within artificial intelligence.

The Overlap of the 8 Categories

When we started our discussion of the categories of AI, we did state that the categories are not mutually exclusive. And, indeed, they are not. Think in this way.

- Artificial Narrow Intelligence - a category that includes the (sub) categories of reactive AI and limited-memory AI.
- Artificial General Intelligence - a category of its own.
- Theory of Mind Artificial Intelligence - a category of its own.
- Self-Aware Artificial Intelligence and Artificial Superintelligence - 2 AI category names describing the same type of AI.
- Generative Artificial Intelligence - a category of AI that may and already does in some instances have a place in all the other categories.

The last, generative AI, is worth a bit more of a discussion. Large language models, specifically, will play an important part as the front end for many other AI tools. LLMs will support natural language processing, so that we can communicate with an AI tool using our natural language. As well, generative AI tools will support the more natural communication process between people and technology. Think about having a simple conversation with a theory of mind AI about the news of the day. The theory of mind AI will use LLMs to carry on a normal conversation with you, generating new content in response to the topic.

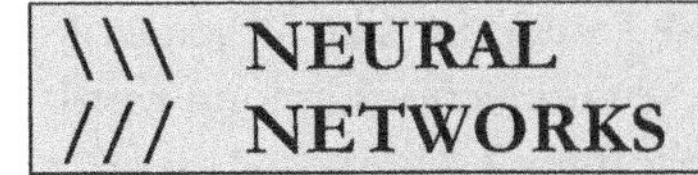

There are several different types of software tools for implementing AI. We previously talked about an expert system, a tool for creating if-then rules. The expert system follows the rules to produce an outcome. In this way, an expert system is a form of a reactive AI. It cannot adapt itself, create new rules, or learn. It simply reacts to the rules and the inputs you provide.

The most common tool for creating limited-memory AI is an artificial neural network. An ***artificial neural network (ANN)*** is a software tool with workings patterned after the human brain, including things like axons, dendrites, neurons, and synaptic connections. (All those wonderful things you learned in your biology class.) While artificial neural network is the correct technical term, you'll often see the term *neural network*.

Neural networks are built in such a way that they can change the manner in which they work, continually updating themselves and becoming better at the decision-making task. That is, neural networks exhibit the human intellectual task of *learning*.

Neural Network Layers - Input, Hidden, Output

Let's consider a simple example, that of your wanting to build an AI that can determine if a photo is that of a dog or a cat. (A cat for sure, not guacamole.) You can do so using a neural network like something in Figure 3.2 on the opposite page.

Notice that a neural network has 3 layers:

1. Input - the data you provide to the neural network
2. Hidden - an interior set of nodes that accept the data from the input layer, mathematically manipulate them, and send a value to the output layer
3. Output - the layer that gives you an answer (i.e., either this is a dog or this is ~~guacamole~~ a cat)

Figure 3.2 **Simple Neural Network**

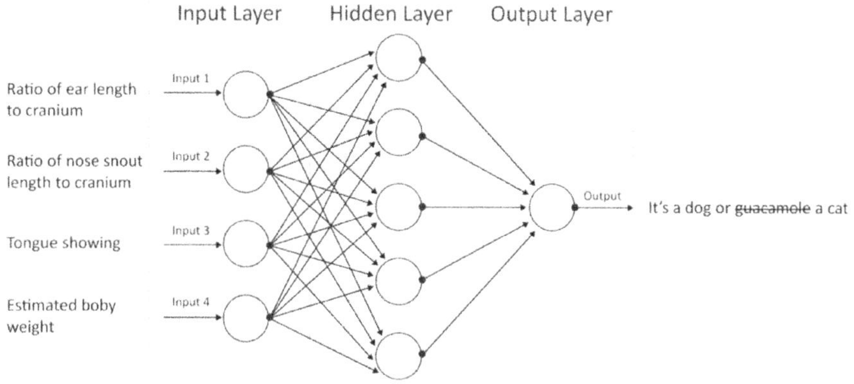

Input Layer Hidden Layer Output Layer

Ratio of ear length to cranium — Input 1

Ratio of nose snout length to cranium — Input 2

Tongue showing — Input 3

Estimated boby weight — Input 4

Output — It's a dog or ~~guacamole~~ a cat

So, you build some image processing software that can look at a photo and gather the following inputs:

- Ratio of ear length to cranium (dogs usually have longer ears in relation to their head size)
- Ratio of nose snout length to cranium (dogs usually have longer nose snouts in relation to their head size)
- Tongue showing (dogs show their tongues more than cats do)
- Estimated body weight (dogs are usually bigger than cats)

Each input node accepts its specific data from the image processing software and passes it on to the interior or hidden nodes. In the hidden nodes is a formula that basically takes the input values, multiplies them by some weights, adds all the results together, and passes the result to the output node.

The output node then sums all the value it receives from the hidden nodes. If the summed value is greater than a threshold value, then the neural network will tell you that the photo is that of a dog. If the summed value is less than the threshold value, then the neural network will tell you that the photo is that of a cat.

Presto. You have a neural network that can distinguish between a cat and a dog. Of course, what's remarkable about your neural network is its ability to *learn*. Let's take a look.

Machine Learning

Machine learning describes the ability of an AI to learn from data by changing the way it works. Machine learning gives AI the ability to get better over time, making better decisions based on how it just performed. Using machine learning, you feed your neural network hundreds and hundreds (perhaps even thousands) of photos of both cats and dogs. After you feed in a photo and your neural network "takes a guess," you tell the neural network whether it was right or wrong. This is training.

If wrong, the neural network goes back into the hidden layer of nodes and ever so slightly changes one or more of the weights in those nodes to get a slightly different overall value in the output node such that the output node overall value is closer to being on the other side of the threshold value.

If right, the neural network goes back into the hidden layer of nodes and ever so slightly changes one or more of the weights in those nodes to a get a slightly different overall value in the output node such that the output node value is even farther away from the threshold value (but on the same side of the threshold value).

Either way, this is the concept of *reinforced learning.*

As you can see, the neural network is learning. With every new photo and either a right or wrong answer, it is changing the weights in the hidden nodes to get closer to the right answer (if it guessed wrong) or positively reinforce a right answer. So, the neural network is literally changing how it works by adjusting the weights.

Deep Learning

If you actually implemented our dog versus cat AI, the AI would get pretty good over time, getting better at distinguishing between a dog and cat with each new photo. It may get to a point where it can perform the task correctly 80 or 90% of the time. To get even more accurate, our neural network would need deep learning.

Deep learning is a subset of machine learning and the ability of an AI to get really, really, really good at making a decision by increasing the number of hidden layers, as in Figure 3.3 on the opposite page. (All deep learning is machine learning, but not all machine learning is deep learning.)

Figure 3.3 **Multiple Hidden Layer Neural Network**

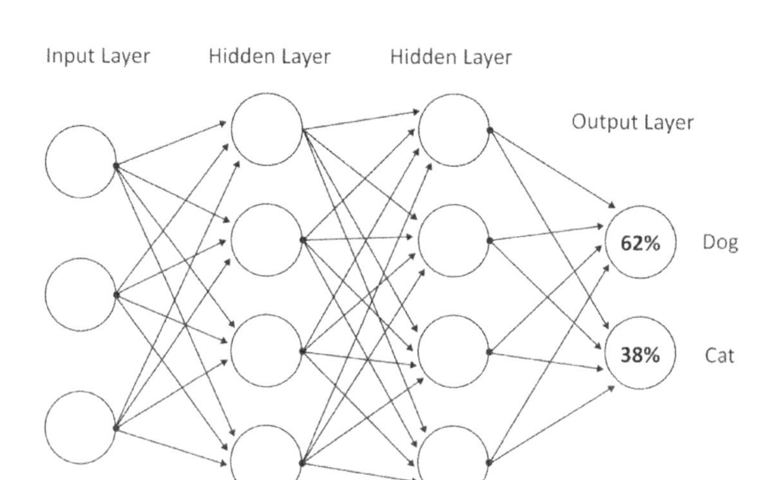

Without getting into too much technical detail, the more hidden layers you have, the more the neural network can take into account and work with subtleties and nuances in the data. It's rather like adding another layer to a decision-making process.

So, deep learning can help create a more effective AI. But deep learning does require much more learning data and a longer learning process. If you double the number of nodes in the hidden layer(s), it makes sense that more training will be required to get the weights set appropriately.

As long as we're talking about dogs, a great example of the need for deep learning would be to build an AI that can distinguish between all the different breeds of dogs. Do you know how many there are? Well, depending on who you ask, there are somewhere between 190 and 360. (Weird… we can't even come up with an exact number. Damn those labradoodles.) To distinguish between all of these, we would need a deep learning AI to work with all the subtleties of the very slight differences among so many breeds.

Supervised and Unsupervised Learning

In our dog/cat example, we used supervised learning. ***Supervised learning***, in artificial intelligence, is a training process in which you specify the inputs to be used and then also identify whether an outcome is right or wrong. Again, in our example, we constrained the AI to just using the 4 inputs we specified. But there are most likely other characteristics that may help distinguish between cats and dogs… paw size, sharpness of nails, something about the tail, perhaps something about fur… you get the idea.

So, we could have opted for ***unsupervised learning***, a training process in artificial intelligence in which you do not specify the inputs and you simply let the AI determine the best sets of inputs to use to arrive at a correct answer.

For example, our dog/cat AI may start to take into account the setting of the photo, either indoor or outdoor. Just guessing here, but if you took all the photos of cats and dogs on the Internet, we bet the biggest proportion of dog photos leans more toward an outdoor setting, while the biggest proportion of cat photos leans more toward an indoor setting.

Collar presence may be another. If you take all the photos of cats and dogs on the Internet, what do you think? More dog photos with collars or more cat photos with collars? Interesting.

It's really difficult to explain unsupervised learning with just words on a page. So, we recommend YouTube and https://www.youtube.com/watch?v=qv6UVOQ0F44&t=283s for a great illustration. In this video, you can watch how an unsupervised neural network learned to master Super Mario World in just 24 hours. Truly, truly fascinating to see how a neural network mastered the game with only the single and simple instruction of maximizing fitness.

Interpreting Images, Video, and Audio: A Convolutional Neural Network

Neural networks are being increasingly used to interpret images, videos, and audio signals. For example, driving systems in autonomous vehicles use neural networks to distinguish among images such as animals, people, street signs, structures (bridges, walkways, light poles), etc.

But working with images (and audio) is different from working with numbers (how much money did you make last year) and categories (in what country were you born). In fact, the process of interpreting images via a neural network is much more complicated. Each pixel in an image is a single input, and some images have millions of pixels. This greatly complicates how image data is fed into a neural network.

To work with images, videos, and audio signals, we use a convolutional neural network. A ***convolutional neural network (CNN)*** is an artificial neural network that is specifically designed to handle the complexities of working with images, video, and audio signals by optimizing the manner in which the inputs are received and processed.

\\\ LEARNING FROM /// MOTHER NATURE

To facilitate the learning process - either supervised or unsupervised - we've *learned* from mother nature, the most sophisticated natural learning intelligence to ever exist. Our "mother nature" concepts include genetic algorithms, neuroevolution, and biomimicry.

Genetic algorithms are a set of techniques modeled after biologically inspired intelligences that primarily include the following:

- Selection - choosing a better outcome over a poorer outcome
- Crossover - combining 2 "good" outcomes (or elements of 2 good outcomes) to see if a better outcome can be achieved
- Mutation - randomly trying something new to determine if a better outcome can be achieved

The thinking is quite simple. In selection, a better outcome is chosen over a less optimal outcome. In crossover, an attempt is made to combine two good outcomes to create a better outcome. In mutation, just try something "new" to see if you get better results. You've probably done each of these while playing a video game. You've opted for one strategy over another because the first yields better results (selection). You've combined the simultaneous pressing of 2 buttons to see if you get a better result (crossover). And you've probably gotten stuck and just randomly tried something new to see if you can get "unstuck" (mutation).

These are all concepts we can implement in artificial intelligence.

Neuroevolution is the use of genetic algorithms within an unsupervised learning context. With neuroevolution, you allow techniques like selection, crossover, and mutation to be the primary drivers of how a neural network learns without specifying the exact inputs, or potentially even the outputs to a given set of inputs. Check out the Super Mario example on YouTube we mentioned earlier.

Biomimicry is the simulation and use of elements of nature to solve human problems. For example, we've learned much from hiving insects (bees, wasps, etc.) to determine how to better construct things like buildings and homes. We've studied forager ants to determine how to better design supply chain management activities. We've studied birds to build better the aerodynamic nature of airplanes. We studied termite mounds to better understand air flow and ventilation.

Fascinating field. You should do some further research into the study of biomimicry and its applications.

\\\ BUILDING YOUR /// OWN AI

You can build your own AI, although we will tell you artificial intelligence development is among the most difficult to do. You'll most likely need to learn a specialized programming tool that supports AI development. Below are some recommendations. [5]

- Scikit-Learn - https://scikit-learn.org/stable/
- TensorFlow - https://www.tensorflow.org/
- Caffe - https://caffe.berkeleyvision.org/
- Theano - https://pypi.org/project/Theano/
- Keras - https://keras.io/
- MXNET - https://mxnet.apache.org/versions/1.9.1/
- Microsoft CNTK - https://learn.microsoft.com/en-us/cognitive-toolkit/

- Google AutoML - https://cloud.google.com/automl
- Accord.NET Framework - http://accord-framework.net/
- Torch/PyTorch - https://pytorch.org/

In your review of these, you'll notice that many are Python-based. So, we definitely recommend that you have some Python experience. (Python is also the most in-demand programming language by employers.)

\\\ BIASES IN /// ARTIFICIAL INTELLIGENCE

Unfortunately, you can build biases into an AI without even realizing it.

Think about building an AI to determine something like credit approval, amount of a loan, or some other financially based decision. You thoughtfully decide to eliminate inputs around demographics… age, gender, race, and ethnicity. And instead, you choose to include education level, length of employment, salary, and amount of savings.

Your thinking is simple. A better credit risk is someone with a higher level of education, a longer length of employment, a higher salary, and a higher amount of savings. That does make sense. But who have you just described? To be blunt and completely honest, you've described the typical white middle-aged male. Without explicitly programming demographic inputs, you have implicitly described a specific group of people, a group of people who have benefited from their demographic status and, as a result, typically have a high level of education, more consistent employment, a comparatively higher salary, and a comparatively higher amount of savings.

Even if you eliminate demographic inputs, you must understand that demographic inputs often have a direct correlation to other (often behavioral) inputs.

This is a hugely important issue we must address in the development and use of artificial intelligence.

\\\ APPLICATIONS OF /// ARTIFICIAL INTELLIGENCE

Okay, there are way too many of these to list here. AI, specifically as artificial narrow intelligence, has been around for some time and is very successful. Let's cover a few of those here for you. [6]

- Roomba, by iRobot - Roomba uses AI to scan the room for furniture and other types of obstacles, determine the room size, and build a memory of the most efficient path for cleaning a room.
- Olly, by Emotech - Olly is an AI assistant like Alexis but has an evolving personality (i.e., embracing theory of mind characteristics).
- Covera Health - uses AI to reduce the number of misdiagnosed patients. The AI sifts through previous patient histories to provide a better profile of symptom information.
- Autonomous vehicles - duh. (Read Chapter 6.)
- Betterment - an AI financial advising platform. It uses AI coupled with the profile of an investor and a wide array of financial investment alternatives to recommend an investment portfolio tailored to the investor.
- Hopper - uses AI to help predict when you, as a consumer, can get the lowest prices on flights, hotels, and other travel-related expenses.
- Social media - from Meta to X to Slack, all social media platforms are running AI to better match people and content.
- Grammarly - uses AI to improve your writing.
- Smart thermostats - use AI to learn occupants' behaviors and adjust the thermostat accordingly.
- KenSci - has built an AI platform to help identify fraudulent insurance claims.
- Biometrics - many applications of AI in biometrics including facial recognition.
- Supply chain management - all the big delivery players - Amazon, UPS, USPS, FedEx, DHL, etc. - are using AI to determine the optimal routes for delivering packages, how to optimize the loading of trucks, and many other aspects of delivery.

The Web 3.0 isn't specific to or a subset of artificial intelligence. Which begs the question, "Why, then, is it in the chapter on AI?"

In formulaic terms, the Web 3.0 looks like this.

Web 3.0 = Web + AI + DLT + IoT
Where DLT includes:
Blockchain
Cryptocurrency
Tokenization
Smart Contracts
Dapps
DeFi

The Evolving Web

Web 1.0, the first iteration of the Web, was characterized by few creators of content (usually large organizations) and many users or consumers of content. Web sites and pages were mostly static (meaning their content often had to be updated manually by changing the underlying HTML code). We're well beyond Web 1.0.

Web 2.0, the current iteration of the Web, is characterized by massive amounts of user-generated content (blogs, reviews, social media, etc.), with all content being dynamically delivered and updated in real-time, often responding to actions and requests initiated by users. Web 2.0 is often called the *participative social web*. But "social" is just one aspect of the dynamic nature of the Web 2.0. We can watch sports updates in real time, changing bids on eBay, etc. Basically, we can see content on the Web today as that content emerges, changes, and eventually dies off.

Web 3.0 is the next evolution beyond Web 2.0 that will exhibit semantic characteristics (AI), decentralized protocols (distributed ledger technology), and be ubiquitous (Internet of All Things). Of course, now we need to chat about semantic, decentralized protocols, and ubiquity.

Semantic

Semantic refers to a concept called *metadata*, essentially data about data. Using metadata, Web 3.0 applications will be able to - in automated fashion - connect related data sources from all over the world. This will require the use of artificial intelligence to make sense of and interpret the data. Consider these:

- I love Bitcoin
- I <3 Bitcoin
- I heart Bitcoin
- I ♥ Bitcoin

Syntax and use of words and symbols are different; the semantic meaning of them is essentially the same. We can understand that; computers will need AI to do so.

Decentralized Protocols
As we discussed in Chapter 2, most information storage and processing are still centralized. The Web 3.0 requires the decentralization of information and software using distributed ledger technologies like blockchain, cryptocurrency, tokens such as NFTs, smart contracts and Dapps that automate tasks, and decentralized financial (DeFi) applications.

Ubiquity
Ubiquity simply means everywhere. For the Web to be everywhere, everything must be connected to the Web. Which means that the Internet of All Things will play an important part in the foundation of the Web 3.0. Connecting all things - computer-related, electronic, and non-electronic - to the Web will create a Web 3.0 that is everywhere.

We're definitely not at Web 3.0 yet. But we'll eventually get there in the 4th industrial revolution. As we evolve to Web 3.0, you probably won't even notice the changes as they occur. They will be subtle and small. But, over the next 10 years, the Web will change dramatically, with all those subtle and small changes accumulating into a significant shift in the Web.

\\\ THE END GAME: /// WHAT THE FUTURE MIGHT HOLD

In the Next 10 Years, You'll Work Alongside an AI
We firmly believe that most all jobs in the next 10 years will involve working with an AI. And we're not talking about just "professional" jobs like financial managers, product planners, marketing mix managers, and engineers. We're also talking about people who work the production floor in a manufacturing environment… all of us (more appropriately, all of you) will have an AI as a colleague, in some form or fashion.

You'll Need a Course in AI Resource Management
Right now, most every business program in the country includes a course in human resource management, how to effectively manage people to maximize their productivity, enable personal and professional development, resolve conflicts, etc. Future business programs will include a course in AI resource management.

An AI may very well come to act like people. It may have a bad day. It may not want to come to work. It may take a mental health day. It may be distracted by world events.

Those statements may seem far-fetched, but don't discount them too much. Theory of mind AI progress is at a feverish pace.

Add Just a Pinch of AI
As we discussed in Chapter 1 and the Internet of All Things, we'll see AI in most all settings. Bathtubs that stop filling when the water level gets too high… a simple reactive AI. A bathtub that stops filling when the water level gets to where you like it… a simple limited-memory AI. Books that automagically generate more problems for you to work because you're struggling with a particular topic… a generative AI.

To start innovating in this space, go back to "add just a pinch of…" The pinch in this instance is AI, obviously. To what existing products and services can you add some AI? What would be the benefit? How would the AI get trained?

And the Oscar Goes to… an AI
There has been much uproar in the arts space (music, movies, etc.) recently about the role of AI in "creativity." Screenwriters fear losing their jobs to an AI. AI have won art competitions. AI have won writing competitions.

AI's role in creativity is going to be interesting to watch.

Job Loss and Gain

Throughout the 275 years of industrialization, there have always been concerns that new technological advances will create unemployment. Partially true to be sure. An IBM survey in 2019 estimated that 120 million workers would need retraining in the following 3 years because of AI's impact on jobs. [7]

But each technology invention and revolution has created more jobs and increased our quality of life. No doubt, many jobs will be lost to AI, but even more will be created. We've been through this many times before. The telephone and telegraph ended the employment of the riders for the Pony Express. Advances in communications technologies and software displaced most telephone operators. But, in each instance, more new jobs were created.

Our willingness to invest in retraining people and workers is a must.

CHAPTER 4

Extended Reality

No Matter Where You Don't Go, There You Are

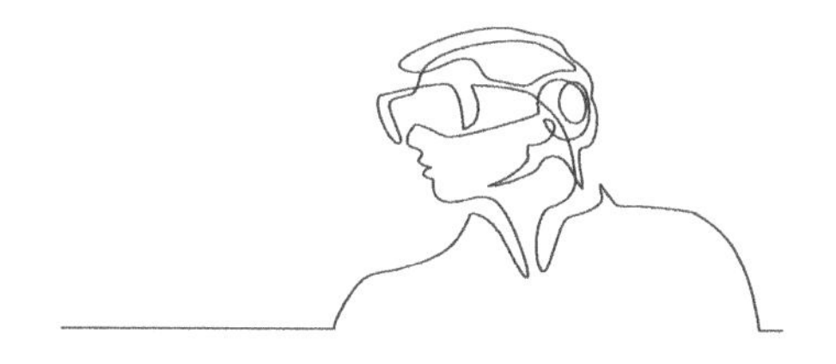

Extended reality is an umbrella term for augmented reality, mixed reality, and virtual reality, which are all some combination or meshing of the physical and virtual worlds through technology.

Consider a continuum from the completely physical to the complete virtual (See Figure 4.1). On the left side is the completely physical world, with no technology enhancement. That's basically experiencing life without the aid of technology. (If you haven't recently, you should try it. The purely physical world is wonderful.) On the right side is a completely virtual world, what is commonly called virtual reality, something you've most likely experienced.

Figure 4.1 **The Physical to Virtual Continuum**

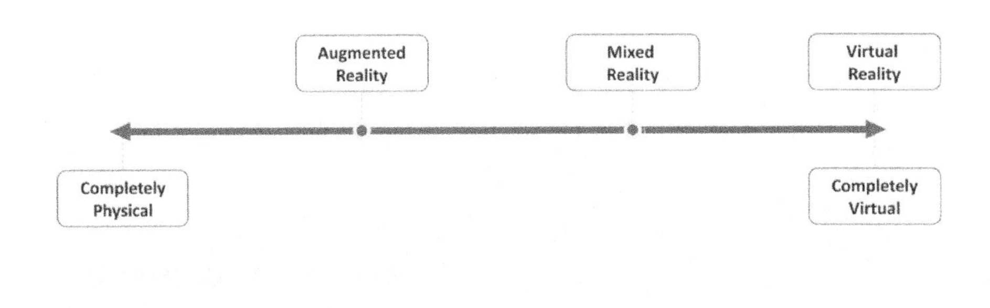

In between the completely physical and virtual worlds are augmented and mixed reality. Augmented reality is based mostly in the physical world with technology-enhanced content and information. Mixed reality is the combining of the physical and virtual worlds in which the two must intersect, be constrained by each other, and "fit" together (i.e., play together nicely in the same sandbox).

\\\ AUGMENTED /// REALITY

Augmented reality (AR) is the enhancing of your view of the physical world by adding, content, information, and the like.

Augmented reality is already very much a part of your life, for example, while watching sports on TV. If you're watching Olympic swimming, you'll see added lines that mark the world record and Olympic record time splits. If you're watching (American) football, you'll see lines that show the distance to a first down or where the team needs to get for a field goal. Replays of soccer goals will show ball trajectory and speed. The list goes on and on.

So, augmented reality is about taking an image (video or photo) of the physical world and adding content to it to enhance, extend, or *augment* what you're seeing.

Your generation played Pokemon Go, which is augmented reality based. Go to the appropriate location and use your phone to see and capture characters. Many restaurants now have AR-enhanced menus. Use your phone to view a menu item and content will pop up showing the ingredients, how it's prepared, and so on. Google Translate can translate (duh) content, text, and road signs as you view them through your camera.

So, augmented reality exists mostly in the physical world, with added (virtual) content to enhance or augment what you're seeing.

What Technologies You Need
You don't really need any "special" technology to take advantage of almost all augmented reality applications, just your camera or a viewing screen like your TV.

Not much more to say.

Building Augmented Reality Applications
If you're interested in learning how to build augmented reality applications, consider these.

- Unity (www.unity.com) - Unity was developed as a tool for game development, first for iOS and then extended to Android. It was the tool used to create the likes of Pokemon Go and Call of Duty: Mobile. It's become a very popular tool for developing games, augmented reality applications, and virtual reality applications. It's also a great tool if you're just learning to build augmented reality applications, and it's very robust for advanced developers.

- Vuforia (www.vuforia.com) - Vuforia is a cross-platform augmented reality development tool, meaning that you can build for both iOS and Android devices within a single tool. It's integrated with Unity. If you search for courses in Udemy for AR development, you'll find many that teach a combination of Unity and Vuforia.
- ARKit - ARKit is Apple's AR development tool for iOS.
- ARCore - ARCore is Google's development tool for Android.

Obviously, ARKit and ARCore are specific to a technology platform, either iOS or Android. Unity and Vuforia support cross-platform development, so they get our recommendation. You can learn the basics of both in Udemy for about $100 (sale prices can go down to around $20 for a course) in less than 20 hours. Not a bad way to spend a few days.

What are you going to put on your resume… that you know Microsoft Office tools like Word and Excel, or that you know those and can also build augmented reality applications?

Applications of Augmented Reality

There are many great examples out there of augmented reality.

In the app world, take a look at Google Maps, Civilisations AR, Froggipedia, The Machines, Smash Tanks, Big Bang AR, Vuforia Chalk, Thyng, Insight Heart, Houzz, and IKEA Place. [1] The latter 2 let you pick furniture and see how they fit into your home, apartment, or dorm room.

Enterprise AR is going to be big business. *Enterprise AR* focuses on the use of AR for organizations. [2]

- Remote Assistance - helping users troubleshoot problems and perform maintenance and repair (Mercedes-Benz, Vodaphone, and Prince Castle).
- Medical Diagnostics - diagnosing diseases and illnesses (Aris MD, OrcaHealth, and surgeons in Brazil).
- Marketing and Sales - enhancing marketing material, moving through the sales process, providing post-sales support (Go-Digital, Cannondale, and Andersen Corporation).
- Training - not much explanation needed here (AGCO, Japan Airlines, and Siemens).
- Logistics - improving daily operations in warehousing, transportation, inventory management, etc. (Amazon, BMW, and DHL).
- Manufacturing - identifying and resolving issues on the manufacturing floor (VitalEnterprises, Airbus, and Bosch).

- Prototype Design - building and reviewing prototypes of new product designs (Aecom, AirMeasure, and BNBuilders).

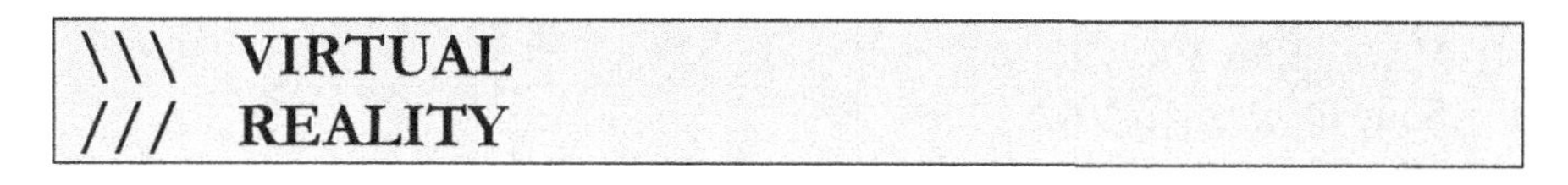

On the other end of the spectrum is virtual reality. ***Virtual reality (VR)*** is a completely immersive simulated virtual experience, which may be based in a real-world environment or a fictitious one. So, you can sit in the comfort of your home and use virtual reality to experience walking the Great Wall of China or swimming the Great Barrier Reef off the coast of Queensland, Australia. You can also experience the wreckage of Jurassic Park (Jurassic Park: Afterworld), or you can leap, slide, run, and jump across city rooftops (Stride). The Great Wall of China and the Great Barrier Reef are based on real world environments. Jurassic Park and Stride's city rooftops are based on fictitious ones.

The focus of virtual reality is to transport you via your senses to another place and set of experiences. Your senses here include what you see, what you hear, and what you feel. (Feeling also works in reverse, as the VR system will detect your movements and respond appropriately.) And, some VR applications are starting to incorporate the sense of smell. Check out OVR Technology at www.overtechnology.com.

What Technologies You Need

Most people your age have experienced virtual reality, as it's a very popular platform for gaming, including Arizona Sunshine, Batman Arkham VR, Astro Robot: Rescue Mission, Beat Saber, Elite Dangerous VR, Fallout 4 VR, Half-Life: Alyx, Lone Echo 2, Minecraft VR, Pistol Whip, Resident Evil 4 VR, and many, many, many others.

To experience VR, at a minimum you'll need a VR headset (an electronic headset that adjusts its field of vision to your moving your head in different directions, up and down and side to side) and some sort of hand controller for selecting menu items, pointing at things, and other interface control tasks. While hand controllers can perform very basic functions like those listed, they can also be used as specialized controllers depending on the situation, for example, a fishing pole, a weapon, a bowling ball, a light saber, etc.

Popular VR headsets include:
- HTC Vive Pro 2
- Meta Quest Pro
- Valve Index VR Kit
- Sony PlayStation VR2
- Meta Quest 3

Of course, these are changing and being updated all the time.

If you haven't yet, we would recommend you try out Google Cardboard. It only costs about $10 and isn't electronic at all. Instead, you drop your phone into the Cardboard and your phone becomes your screen. A very inexpensive way to experience virtual reality.

Building Virtual Reality Applications

You can build your own VR applications, but it's definitely more complicated than building augmented reality applications.

Many of the VR development tools can also help you build augmented reality applications. If you're interested in building VR, we would recommend that you start by building some AR applications first because they're easier. As you get more comfortable with the tool, you can then move on to VR applications. VR development tools include: [3, 4]
- Unity VR Development
- Blender
- Maya
- Unreal Engine
- A-Frame
- VRTK
- Open VR
- Amazon Sumerian
- echoAR
- Eyeware Beam Head and Eye Tracking SDK
- eevo VR Development
- WorldViz Vizard
- Cryengine VR Development

Several of these also pop up in the 3D printing space, which we'll discuss in the next chapter.

Applications of Virtual Reality
As with augmented reality, applications of virtual reality abound.

In the personal space, we've already mentioned games including Arizona Sunshine, Batman Arkham VR, Astro Robot: Rescue Mission, Beat Saber, and many others. Consider also checking out Google Expeditions, Colosse, Titans of Space, Google Earth VR, Kingspray Graffiti VR, Ocean Rift, Foo Show, Virtual Desktop (your computer in VR), and Within.

On the enterprise front, all organizations are racing to take advantage of virtual reality. [5] For each area below, we've included one or a few examples, but there are many others.

- Online Shopping and Retail (IKEA Reality Kitchen Experience for designing your perfect kitchen)
- Engineering and Manufacturing (Airbus for determining the comfort of seats)
- Tourism and Hospitality (virtually experiencing museums, resorts, and travel options)
- Communication and Collaboration (virtual team meetings)
- Trainings (sales, military, and many more)
- Architecture, Construction, and Design (what the interior of a building will look like after building it)
- Medical and Health Care (treating psychological disorders, practicing surgery on virtual cadavers)
- Education (experiencing learning as opposed to listening to a lecture)
- Sports (athletes experiencing simulated training sessions and in-game situations)

In between augmented and virtual reality is mixed reality. ***Mixed reality (MR)*** is the merging of the physical and virtual worlds in which both worlds simultaneously co-exist in real-time and are constrained by each other.

Let's consider a simple example, that of designing your dorm room, focusing on the optimal layout of your furniture. Using mixed reality, you would first use your phone or special MR headset to *map* the room. (Yes, you can do this now with the latest phones that have Lidar capabilities.) In mapping your room, you're getting the dimensions of the room, including things like the height of the window, how far the door is from each wall, how wide the door is, the exact location of the sink (if you're lucky enough to have one), and so on.

You would in similar fashion map the dimensions of the existing furniture. Then, with mixed reality you could virtually move the physical furniture around the room. You could virtually add new furniture, with the mixed reality not allowing you to block the door with a chair or bed, for example. Thus, the coexisting of physical and virtual.

You get the idea. Mixed reality is a combination of the physical world (the dimensions of your room) and the virtual world (the dimensions of the furniture you're thinking about adding), with the physical constraining the virtual and vice-versa.

What Technologies You Need

To experience mixed reality, you can use your phone and some apps like Houzz or IKEA Place to do what we just described in designing your dorm room (or apartment or even an entire home).

More realistically, you'll want a mixed reality headset. A mixed reality headset is similar to a VR headset, except that a mixed reality headset includes the mapping capability of a room or environment. The dominant mixed reality headsets include: [6]

- Apple Vision Pro
- Microsoft HoloLens 2
- Magic Leap 2
- Meta Quest Pro
- Lenovo ThinkReality VRX

- Varjo XR-4 (and XR-3)
- Meta Quest 3

Some of these support both mixed reality and virtual reality applications.

Building Mixed Reality Applications
In the mixed reality space, you'll need to learn development skills particular to the headset hardware you choose to use (those listed above).

Microsoft and Magic Leap dominate the mixed reality space, so consider going with either of those. Apple is a new entrant in this space and should certainly be considered an important player. If you want to pursue a career in mixed reality development, starting salaries are in the $150,000 - $200,000 range. Very nice.

By the way, if you do choose to learn mixed reality development, you'll also be learning augmented and virtual reality development. So, consider perhaps learning basic augmented and virtual reality development first as a lead into what can be very complex development in mixed reality.

Applications of Mixed Reality
Mixed reality is the newest among the three in the extended reality space. To get a feel for mixed reality, we recommend that you watch the following videos:

- Envisioning the Future with Windows Mixed Reality - https://www.youtube.com/watch?v=96tklVB8X8o
- HoloLens 2 AR Headset - https://www.youtube.com/watch?v=uIHPPtPBgHk&t=307s
- Introducing Microsoft Mesh - https://www.youtube.com/watch?v=Jd2GK0qDtRg and https://www.youtube.com/watch?v=_0InCXA13L8
- Meta Quest 3 Mixed Reality Game Play - https://www.youtube.com/watch?v=BM0CZUHrbcw

You can also find examples of mixed reality in:

- Medical education (https://www.youtube.com/watch?v=f6KK4IsorUI)
- Gaming (Star Wars, Game of Thrones, Angry Birds)
- Merchandising (for determining optimal store layout virtually before physically moving everything around, over and over again)
- Urban planning (for understanding the impact of new buildings, bus stops, etc., see, for example, https://www.youtube.com/watch?v=f65Gf_immrs)

One of the big opportunities for mixed reality is in the team collaboration space. We can already have virtual meetings, bringing people together from all over the world. But, with mixed reality, we can actually bring people virtually into the room with avatars and holograms. Using your mixed reality headset, you'll be able to see 3D representations or holograms of your co-workers in the room with you.

\\\ COMPLEXITY OF DEVELOPMENT /// AR, VR, AND MR

In terms of complexity of development, augmented reality is by far the easiest, mixed reality is the most difficult, and virtual reality is somewhere in between (See Figure 4.2). It makes sense when you think about it.

Figure 4.2 **EXTENDED REALITY (XR) – COMPLEXITY OF DEVELOPMENT**

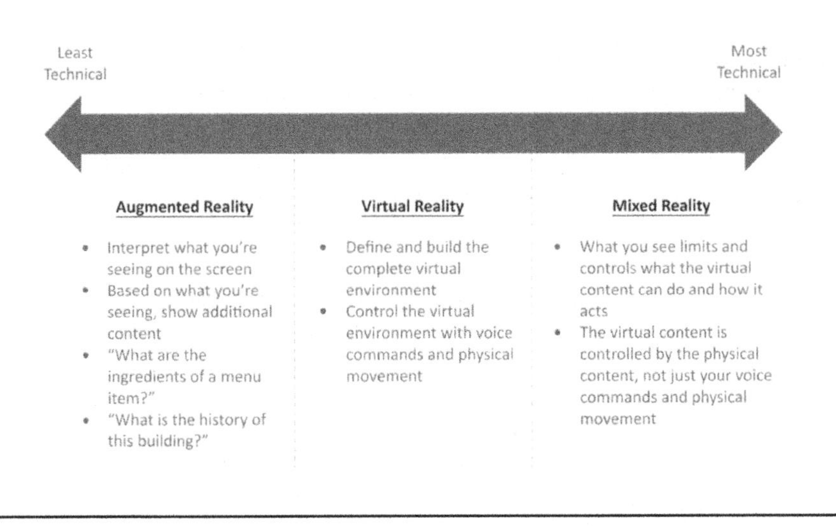

Augmented reality is about recognizing images and the like through a camera and then laying content on top of those images. Building the very basic version of Pokemon Go isn't that difficult. You need to access the location of the user via their phone GPS, track the user to a specific location, present an image of a Pokemon character on the phone's screen, and enable the user to capture the character in some way.

(2023 was an important "anniversary" of sorts for GPS. It was the 50th anniversary of the approval of the GPS program. And it was the 30th anniversary of the GPS program going "live" for public use.)

Virtual reality is more challenging because you have to build a complete virtual world and capture and respond to a lot of movement by the user (looking in different directions, hand movements, etc.). You also have to build the virtual environment on a *movement continuum*. That is, as the user progresses through a virtual environment, you have to have built the virtual environment for the next virtual location. Think about a virtual reality in which you move the user through a building. You have to have all the rooms of the building built (and stairwells, elevators, hallways, etc.) and continually change the view of the building as the user moves through it.

Mixed reality is the most challenging to build because you have to map the physical world around the user and use that mapping to constrain what the user can do in the virtual world. If you're building a mixed reality for someone to virtually place different pieces of furniture in their home, you have to map the size of the room and use those dimensions to ensure that the furniture will fit according to where the user wants to put it.

\\\ METAVERSE – /// EXTENDED REALITY EXTENDED

Metaverse is a virtual universe - aided by technologies such as augmented, virtual, and mixed reality, video, holograms, avatars, and various sensing technologies - in which people connect, live, work, conduct commerce, and play. This can include team collaboration in a business environment, virtual trips, going to conferences and concerts, and even something as simple as having a conversation with an avatar- or hologram-based person anywhere in the world.

Metaverse has gained considerable attention most recently because of Facebook's and Microsoft's big push into the space. In late 2021, Facebook changed its name to Meta Platforms, signaling a significant shift into the metaverse space. According to Mark Zuckerberg, "The next platform and medium will be even more immersive and embodied Internet where you're in the experience, not just looking at it, and we call it the metaverse." [7]

During that same year, Facebook/Meta acquired Within, Unit 2 Games, Bigbox VR, and Downpower. All of those are in the metaverse space in some form or fashion. (Facebook acquired Oculus in 2014.) [8]

At about the same time, Microsoft announced that it was adding HoloLens and its mixed reality capabilities to Microsoft Teams, a collaboration tool that many businesses use. The tool is called Mesh. Mesh allows you to participate in virtual meetings as your actual self or a 2D or 3D avatar or hologram. And you won't need a headset to view other people who choose to go the 3D avatar route. [9]

There are a number of other companies in this space or emerging including:

- Decentraland
- IMVU
- Epic Games
- Roblox
- Minecraft
- Second Life
- Nowhere
- The Sensorium Galaxy
- SuperWorld
- Sandbox
- OpenSea
- Axie Afinity
- Bloktopia
- Spatial

The goal of the metaverse concept is to connect people through a more immersive experience than traditional technology tools like social media and team collaboration tools.

We're going to see entire cities emerge in metaverse. You can already use cryptocurrency to buy NFTs that represent land ownership in metaverse. You can use more cryptocurrency to build on your land, perhaps an amusement park or a concert venue.

You can expect to see "traditional" businesses like car manufacturers setting up shop in metaverse. You'll have an immersive experience test driving a car and determining which features you want. After you've put together your package, you can transfer that to a local dealership and buy the exact car you want.

Even governments are getting into metaverse. In late 2021 the country of Barbados announced it was building an embassy in Decentraland, a popular crypto-powered metaverse. Barbados even declared its embassy land purchase to be sovereign land. Barbados will offer such services as e-visas and the ability to extend cultural diplomacy and the trading of art, music, and culture. [10]

Metaverse is one of the many new "wild, wild wests" in the 4th industrial revolution. There will most certainly be plenty of money made and lost in metaverse. Whether you aspire to be an intrapreneur or entrepreneur, you should consider evaluating metaverse as an investment opportunity.

You Won't Need to Build Extended Reality from Scratch

You'll obviously have to learn to build if you want to pursue a career in extended reality development. But, for the typical consumer or user of extended reality, we're going to have apps and embedded extended reality.

As for extended reality apps, this is going to be no different from today's phone apps. You don't have to write your own text-messaging app; it's already on your phone. You don't have to write your own video-based social media app; you can download it from the app store.

The same will be true for extended reality.

So, we'll have Vapps (virtual reality apps), Mapps (mixed reality apps), and Aapps (augmented reality apps… not real sure how to pronounce *Aapps*, but fortunately pronunciation isn't important when writing a book.)

Phones, tablets, and computers will come with built-in capabilities in the extended reality space. We're already seeing that with the newer phones and tablets that incorporate Lidar and augmented reality enhancement capabilities.

You'll be able to download just about any environment - real or fictitious - for virtual reality. You can almost do that now.

Mixed reality is going to be the really interesting space. We think it becomes a personal productivity tool with built-in libraries or capabilities and images. Think about Microsoft Word (or whatever word processing tool you use). It's a personal productivity tool for creating text-based documents. You can create your resume, an advertising flyer, a term paper, and so on. You don't need different word processing software for each situation.

The same will be true for extended reality. You'll get libraries of images, objects, enhancements, and effects. Your mixed reality headset will map the room or area where you are, and you can start to play, build, work, and interact.

Regarding embedded extended reality, right now, we talk about extended reality as separate from other types of technology. It's new, it's cool, we're trying to figure it out, and we're innovating all the time. Call that Phase #1.

Phase #2 will be what we just described… extended reality as a set of common apps on your technology. You won't need to build the app itself; you'll just build inside the app. Let's call that Phase #2. It will be here shortly.

Beyond is Phase #3. Extended reality will be embedded into every aspect of technology (and our lives). As you watch TV, you'll be able to simply ask, "What's that building?" Your TV will pause the show and talk to you about the name of the building, when it was built, the architect, the height of the building, its location, and so on. You'll even be able to take a virtual tour of the building, before returning to the TV show. You'll be able to do the same when you're taking photos with your phone.

You'll stand in front of a smart mirror trying on clothes and you'll be able to see what you look like in clothes of a different color or with different shoes. Most likely, you won't even have to "try on" clothes. You'll simply select them on the mirror and the mirror will virtually put them on you.

Yes, yes… mirrors are becoming a part of our technology tool set. A ***smart mirror*** is a mirror that displays your image and also extended reality content like augmented information, avatars, changes to your hair style or clothing, and so on. Check out smart mirrors at https://www.chattersource.com/smart-mirror/.

The Blurring of Augmented, Mixed, and Virtual Reality

There is already a lot of confusion and cross over among augmented, mixed, and virtual reality, especially in the popular press. For example, if you search on augmented reality applications, you'll actually find lots of mixed reality examples. Many extended reality headsets support applications for augmented, mixed, and virtual reality. The same is true for many extended reality development tools.

It's not really a big deal; the lines between augmented, mixed, and virtual reality are becoming quickly blurred. Someday we'll simply use the term extended reality.

The Death of Traditional Eyewear

The miniaturization of technology is an important driver, in the past, the present, and will continue to be in the future.

Someday, we won't need today's headsets for enjoying mixed and virtual reality. Capabilities to do so will be built into our eyewear, which will most definitely include sunglasses. We already have smart or AR-enabled ski goggles like RideOn and smart swim goggles like FINIS smart Goggle.

While we may never really see the death of traditional non-technology eyewear completely, it will certainly take a second place to smart eyewear.

The Birth of Real Holograms

Without getting into any level of technical detail, holograms as we know them today are not real holograms. Real holograms are created from a recording of a light field, not a traditional camera lens. Thus, real ***holograms*** are 3D images with depth that have been created using light beams, millions of them. Using panels of lights sitting opposite of each other, two light beams from each panel intersect in space and create a pixel, a tiny dot suspended in space that you can see. You can change the color of the light beams emanating from each panel to create a different color at the intersection. Do this with enough pairs of millions of light beams and you get a real 3D hologram with depth.

Google's hologram initiative is called *Project Starline*. It is a booth-based holographic video conferencing tool. Very cool. Check it out at https://www.youtube.com/watch?v=obuyCkotJ_s and https://www.youtube.com/watch?v=Q13CishCKXY&t=4s.

Real holograms are coming. Hmmm… *real holograms*, another oxymoron?

The Synergies of 4th Industrial Revolution Technologies
The biggest opportunities of the 4th industrial revolution lie at the intersection of the various technologies. Metaverse, for example, uses extended reality such as avatars and holograms, cryptocurrency, and NFTs to build a more realistic and immersive experience.

Even hologram-generating technologies like *Project Starline*, which we just mentioned, make use of artificial intelligence.

As you innovate in the 4th industrial revolution, try to combine multiple emerging 4th industrial revolution technologies. The result is not additive but rather multiplicative. Big, big change, much better world.

CHAPTER 5

3D Printing

One Pair of Shoes for Life

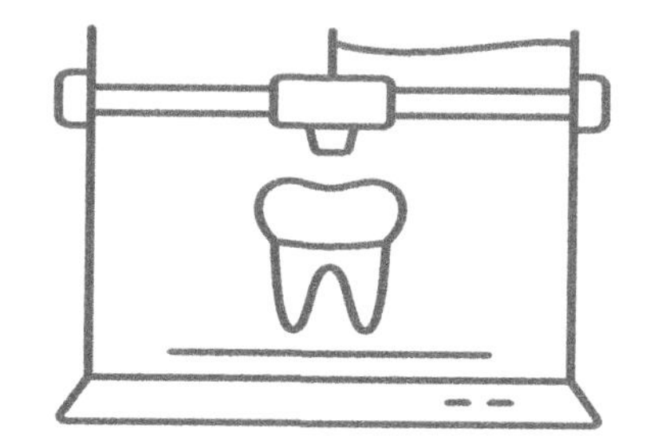

Question: Which of these has not been 3D printed?
A. Toothbrush
B. Doorstop
C. Motorcycle
D. Clothing
E. Boat
F. House
G. Food
H. Lamborghini

We're sure you're getting good at these by now. You're right, all the above have been 3D printed.

Basically, if it's in physical form, you can most likely 3D print it. Now, you may be thinking, "Sure, everything except for living organisms." Think again, more on bioprinting in a moment.

3D printing is a subset of additive manufacturing. And it makes sense that additive manufacturing is the opposite of subtractive manufacturing. ***Subtractive manufacturing*** is simply taking something big and making it smaller. We've been doing some form of this since we invented tools. Early civilizations took a big rock and chiseled it down into a statue, bust, or part of a pillar. Today, we take trees and whittle them down to 2x4s, toothpicks, matches, and baseball bats.

Using the subtractive manufacturing approach, we slice, press, squeeze, cut away, whittle, etc. until we get the finished product we want.

Additive manufacturing is the opposite of subtractive manufacturing and includes things like cast molding, injection molding, and - of course 3D printing. 3D printing has become so synonymous with additive manufacturing that we now use the terms interchangeably. In fact, if you go to Wikipedia and search additive manufacturing, you'll be redirected to the page on 3D printing.

3D printing is the construction of a 3-dimensional object by typically adding (i.e., printing) layer upon layer of liquid material until the object has been completely printed or constructed. The material is initially heated to a necessary liquid state and then printed. Once a layer cools, more material is heated and printed on top of the previous layer. Rinse and repeat until the object is completely printed.

\\\ INTERESTING AND COOL /// INNOVATIONS IN 3D PRINTING

In the opening question for this chapter, we put up a bunch of stuff that has already been 3D printed. The mundane and boring include things like toothbrushes and doorstops. Boats are kind of interesting, check out https://www.youtube.com/watch?v=1uFMCEyAotc. (Hundreds more videos of 3D-printed boats on YouTube.). But the others truly illustrate the power and disruptive impact of 3D printing.

We'll point you toward some videos and resources on each of these. Review them at your leisure.

Motorcycles
European-based BegRep makes a variety of large-scale commercial 3D printers and it also produced NERA, the world's first fully 3D-printed e-motorcycle. It took a short 12 weeks to get from idea to a fully functional motorcycle. The tires, bumper, and seat were all printed with flexible material. (The tires, interestingly enough, are airless.) Even the shock-absorber system was printed with a flexible material, eliminating the need for a complex hydraulic suspension system. Watch the video at https://www.youtube.com/watch?v=s4p59wk_0R4.

One of the great advantages of 3D printing is the ability to customize what you're printing. Imagine being able to design and print a motorcycle or bicycle to fit your exact body style. A seat that fits your posterior perfectly. The handle bars the perfect distance for your arm length. The pedals perfectly placed for your feet and the length of your legs. And you'll get to personalize your motorcycle or bike with your choice of colors, frame style, and accessories like a holder for your IoT-enabled water bottle.

Clothing
It certainly seems that 3D printing is going to be big business in clothing. While we can't yet print purely in fabrics like cotton, we're getting close.

Imagine designing and printing your own clothes at home. It will become a reality. You'll be able to add your own design features to create personalized clothing. You'll be able to tailor your clothing to your unique body style and dimensions. You'll be able to print in whatever colors you want.

Take a look at these initiatives.

- Danit Peleg - truly was a pioneer in the space of 3D-printed clothing. In 2018, she was named by *Forbes* as one of Europe's 50 most influential women in tech. Check her out at https://danitpeleg.com and also https://www.youtube.com/watch?v=3s94mIhCyt4&t=15s. Danit is even selling NFTs, allowing anyone to collect and trade digital fashion. Stepping further into the 4th industrial revolution, Danit uses augmented reality-based body scanning apps like nettelo to accurately measure a person's body dimensions.
- Ministry of Supply - has developed a 3D printing process for knits, specifically blazers for both men and women. Using its technique, Ministry of Supply has noted a 35% reduction in material usage. [1]
- Julia Daviy - quoting her, "Filament is not ready to replace fabric completely just yet, but it's only a matter of time. As it stands today, the technology is already good enough to create better clothes than certain materials, like leather. For example, I created a top and a skirt that looks as if I used a laser to intricately cut them from a piece of leather, but it's entirely flexible and biodegradable, vegetable based plastic. It was faster, cheaper, and more sustainable than using leather." [2] See what Julia is doing at https://juliadaviy.com.
- Viptie 3D - 3D-printed neckties and bow ties for men. [3]
- Shoes - many in this space including Zellerfeld at https://www.zellerfeld.com/ and Futurecraft 4D (an initiative by Adidas and Carbon) at https://www.adidas.com/us/futurecraft.
- 3D-printed jewelry - lots of great stuff going on in this space. See, for example, Ola Jewelry at https://www.olajewelry.com/.
- Eyewear - again, many great examples in this space including Stone 3D (https://stones3d.com/), Monoqool (https://monoqool.com/), and Your Eyewear (https://www.youreyewear.com/).

Stopping here with the list but you should explore more on your own. It really doesn't matter the gender, the type of clothing, shoes, accessories, or even eyewear. 3D printing has a place in all aspects of clothing and the fashion industry.

Think about reducing waste, creating clothing that is biodegradable, and addressing the big hairy audacious problem of fast fashion. We hope you're up for the challenge.

Lots of great opportunities in this space.

Houses

Homes are expensive, time-consuming to build, and require significant energy and resources.

But a 3D-printed home can be quickly built at a fraction of the cost of a traditionally built home. And many 3D-printed homes are using waste and recycled material, making them eco-friendly.

- Chinese company WinSun has made a number of achievements in this area. It 3D printed a house in just one day. It also 3D printed a 5-story apartment complex and an 11,840 square-foot villa. All these printing efforts use industrial waste, recycled construction material, and cement. [4]
- The United States Marine Corps 3D printed a 500 square foot barracks made entirely of cement in just 40 hours. Normally, it would take 5 marines a full week to build the equivalent barracks using wood. [5]
- Construction company ICON has 3D printed homes all over the world. One of its efforts was to 3D print homes in the Austin, TX area, which are now for sale. The homes range in size from two-bedroom to four-bedroom models. Each home took approximately 5 to 7 days to print. [6]
- Italian-based Mario Cucinella Architects, in conjunction with Italian additive manufacturing company WASP, is 3D printing homes with the primary source of material being the surrounding soil, coupled with water and rice husks. These 645 square-foot homes take about 200 hours to print and are dome shaped. [7]

That's just a few of the many initiatives for 3D printing homes. They're affordable, take minimal to construct, and are very environmentally friendly when using recycled material, waste, and soil.

As with 3D printing clothing, the 3D printing of homes can help address an important problem, that of safe and affordable housing.

Food

The 3D printing of food will be fascinating to watch. Most people have no problem with 3D-printed clothing, homes, motorcycles, and the like. But when it comes to eating something that has been printed, people start to hesitate. (Which is really odd considering that popular treats like Gummy Bears come from a form of injection molding, an additive manufacturing technology.)

But, 3D-printed foods, especially those that we 3D print in our homes, will most likely become a part of our everyday lives. You'll be able to add the vitamins, nutrients, and supplements you want. You'll be able to custom print mild, medium, or hot buffalo wings. Parents will be able to use vegetable-based materials and add spices to 3D-printed great-tasting, healthy snacks for their children.

Consider these examples of 3D printing food. [8]

- Food Ink, a popular restaurant in London, 3D prints its entire food and dessert menu. Taking it one step further, Food Ink also 3D prints its chairs, lamps, decorations, and silverware.
- 3D-printing company Beehex provides the technologies for 3D printing pizzas and cookies. (NASA considered using this technology to make pizza for astronauts.)
- Open-Meals is working to 3D print meals specifically designed for the nutritional needs of each patron. One of Open-Meal's initiatives is Sushi Singularity. When making a reservation, you provide biometric samples of yourself including your DNA. Sushi Singularity will use that information to 3D print a meal tailored to your nutritional requirements.
- Novameat and Redefine Meat are 3D printing plant-based versions of meat products. They even use plant compounds that taste like the blood, fat, and muscle in traditional meats. Once you buy them, just throw them on the grill and cook to your liking.
- Netherlands-based Upprinting is collecting food typically wasted because of over-ripeness or "ugliness." It purees the food to 3D print biscuits.
- LA-based Sugar Labs describes itself as, "rogue chefs, architects-turned-designers, and tech geeks" who 3D print sugar-based treats and desserts.

Again, very short list of probably hundreds of initiatives in the 3D-printed food space.

Want to try it at home? Consider these food 3D printers. [9]

- Mycusini, for 3D printing chocolate (www.mycusini.com)
- PancakeBot, a successful kickstarter campaign (www.pancakebot.com)
- Felix Food 3D Printer, chocolate/pasta/puree/meat (https://www.felixprinters.com/felix-food/)
- byFlow Focus, chocolate and thick pastes (https://www.3dbyflow.com/)

Lamborghini

Okay, we had to include this one because it involves a 12-year-old young man, his dad, and - of course - a Lamborghini. Xander Backus and his dad, Sterling, have been working for years to 3D print an exact working, road-ready replica of a Lamborghini. While you may think that Lamborghini doesn't want anyone "printing" its cars, Lamborghini has actually supported the father-son team and provided a 2-week loaner of a Lamborghini Aventador S. Check out one of the many videos about this initiative at https://www.youtube.com/watch?v=c58s1mYmkKw&t=6s.

Bioprinting

Bioprinting (or ***3D bioprinting***) is the use of 3D printing technologies, combined with special filament such as bioink and other biomaterials, to replicate parts that imitate bones, natural tissues, tendons and ligaments, skin, blood vessels, and even organs.

Let's talk about 3D-printed synthetic living skin in burn treatments as an example.

More than 11 million people annually require burn-related medical procedures to improve the functional and aesthetic outcomes of burn wounds. Traditionally, burn treatments have included primary closure, burn wound excision with subsequent skin grafts, and skin substitutes. [10]

Now, medical professionals can 3D print synthetic living skin tissues. The print process combines the creation of a *scaffold* (polymeric biomaterials that provide the structural support for cell attachment and subsequent tissue development) with the depositing (3D printing) of living cells within the scaffold. The bioprinted tissue is allowed to mature and then used as a skin graft of sorts on the burn victim.

Other 3D bioprinting research and applications include:

- Breast implants [11]
- Bone tissue [12]
- Ears and noses [13]
- Blood vessels, nerves, and muscular tissue [14]
- Complex bone tissue [15]
- Teeth and dentures [16]

Related to 3D bioprinting are initiatives like e-BABLE, a global community of "Digital Humanitarian" volunteers who are working to 3D print upper limb prosthetic devices such as hands. The community is open-source, so anyone can use, modify, and reprint a design created by someone else. [17]

3D bioprinting is still very much in its infancy stages of research and application. But considerable progress is being made daily.

You should begin to think about the synergies of combining multiple 4th industrial revolution technologies. Consider 3D printing and IoT. In 2018, researchers at the University of Minnesota successfully proved that they could 3D print biological sensors directly onto the skin of a human being. Fascinating application at the intersection of two 4th industrial revolution technologies. [18]

\\\ MATERIALS IN 3D PRINTING – /// FILAMENT AND RESIN

You can 3D print with just about any material or simulated material you want. Depending on your choice of different type of 3D printer, you'll either be printing with ***filament*** (a solid spool of material, think really thick fishing line) or ***resin*** (a tank of liquid material). Filament and resin are essentially the same; they differ in their starting form (filament is solid while resin is already liquid) and the process of heating and printing them.

Either way, the common materials include:

- PLA (polylactic acid) - the most common for in-home printing, the printing of prototypes, etc. Very inexpensive but also not of high quality or strength.
- ABS (acrylonitrile butadiene styrene) - the second most commonly used, and not much different from PLA.
- ASA (acrylonitrile styrene acrylate) - tougher and stronger version of ABS, making it better for outdoor applications.
- PET (polyethylene terephthalate) - common clear plastic like in water bottles.
- Nylon - synthetic polymer; very strong and durable.
- TPE (thermoplastic elastomer) - think flexibility like in the soles of running shoes.
- PET-G or PETG (polyethylene terephthalate glycol) - clear, chemical-resistant plastic like that of food storage containers. Also, 100% recyclable.
- PC (polycarbonate) - the strongest; think bicycle helmets.

You can also use other filaments/resins, what we call "exotics." Some of those include:

- Hemp
- Wood
- Metal
- Biodegradable
- Conductive
- Glow in the dark
- Magnetic
- Ceramic
- PVA (polyvinyl alcohol) (dissolves in water, the clear plastic material around a laundry detergent or dish washer pods)
- Wax

If you go the filament route, materials come on a spool, about 6 inches in diameter and weighing a couple of pounds. Prices range from $20 per spool for the cheaper filaments (PLA, ABS, etc.) up to $50+ for the exotics. If you go the resin route, you buy resin by the liter, with cheaper resins costing about $40. When you get a liter of resin, you pour it into a tank that your 3D printer uses. The tank itself costs about $80 and you'll have to replace the tank periodically.

\\\ THE 3D PRINT /// PROCESS

The 3D printing process includes 4 steps:
1. Ideate and design on paper
2. Use 3D modeling software to create the digital model
3. Use slicing software to create the instructions your 3D printer needs to print the object
4. Print the object using your 3D printer

Of course, this is a very iterative process. So, when you're first starting on a new object, you should print at least the first iteration using a filament/resin like PLA or ABS, the cheapest of the materials. You'll undoubtedly print your first try at the object and then decide to make some design changes. Using PLA or ABS to get to a final design is the cheapest and often the fastest way to go (the more exotic the material, the longer it usually takes to print).

In the first step, take some time to design your object on paper, thinking about the various dimensions, curvatures, etc. If you want to 3D print a copy of an object you already have (either exact replica or modified version of it), you can take a photo of it and load it into your modeling software.

In the second step, you'll use 3D modeling software to create a 3D CAD (computer-aided design) digital model of your object. The end result of this step is to produce a file called an ***STL file*** (***Standard Tessellation Language*** or ***StereoLithography File***), a digital file that describes the surface geometry of your object. It's rather like using Word to create a .doc or .docx file.

This is the step in which you'll spend most of your time. You can start with a blank slate and use your 3D modeling software to create your model, you can start with a photo as we just mentioned, or you can visit any number of sites and download an STL file of the object you want to print. If you go the latter route, you can modify the design to fit your needs.

There are some really great sites that offer free 3D CAD designs, including [19]:

- Cult (www.cults3d.com)
- Free3D (www.free3d.com)
- GrabCAD (www.grabcad.com)
- MyMiniFactory (www.myminifactory.com)
- Pinshape (www.pinshape.com)
- STLFinder (www.stlfinder.com)
- Sketchfab (www.sketchfab.com)
- Thingiverse (www.thingiverse.com)

Many of these support a very active community of 3D designers with whom you can collaborate on designs.

As for the 3D modeling software itself, there are many options, and some have free personal licenses and/or free licenses for students. There is an important consideration in choosing your 3D modeling software. 3D modeling software is used for a variety of purposes, not solely for 3D printing. For example, you can use 3D modeling software for creating computer graphics for applications like games, virtual reality, and mixed reality. 3D modeling software for 3D printing is often referred to as *solid modeling software.*

You need solid modeling software that creates designs that are referred to as *manifold* or *water tight.* With a manifold or water tight model, the walls have some thickness, and that is necessary for 3D printing. Some 3D modeling software creates walls that have zero thickness, the types of walls that are used in computer graphics and games. This type of 3D modeling software is called *polygon modeling software.* If you use polygon modeling software, you can still create 3D-printable STL files, but it takes a few more steps.

Below are some modeling software options for 3D printing. [20]

- Tinkercad (www.tinkercad.com)
- Blender (www.blender.org)
- BRL-CAD (www.brlcad.org)
- FreeCAD (www.freecadweb.org)
- OpenSCAD (www.openscad.org)
- Wings3D (www.wings3d.com)
- 3D Slash (www.3dslash.net)
- SketchUp (www.sketchup.com)
- Fusion 360 (www.autodesk.com)
- MoI (Moment of Inspiration) (www.moi3d.com)

- Rhino3D (www.rhino3d.com)
- Cinema 4D (www.maxon.net)
- SolidWorks (www.solidworks.com)
- Maya (www.autodesk.com)
- 3ds Max (www.autodesk.com)
- Inventor (www.autodesk.com)

If you want a recommendation, start with Tinkercad, Blender, 3D Slash, or SketchUp. These are very easy tools to learn and offer a free version.

If you love this space and want to pursue a career in the 3D modeling (and printing) space, you should really consider the suite of 3D modeling tools offered by Autodesk, including Fusion 360, Maya, 3ds Max, and Inventor. Fusion 360 is a powerful, general-purpose 3D modeling tool that is very popular in the industry. Maya is great for developing characters. 3ds Max is great for games and the entertainment space. Inventor is great for mechanical design. All these come from Autodesk, so the interface is similar.

Once you've built your model and resulting STL file, you've got one more step to go before 3D printing. In this 3rd step, you need to use slicing software. ***Slicing software*** takes an STL file (the surface geometry of your object) and slices it to create the layers needed for 3D printing. The result is ***g-code***, a CNC (computer numerical control) set of instructions that basically tells your 3D printer how to print your object layer by layer.

Again, lots of options here, including [21]:
- Cura (www.ultimaker.com)
- Simplicity 3D (www.simplify3d.com)
- Slic3r (www.slic3r.org)
- Repetier (www.repetier.com)
- KISSlicer (www.kisslicer.com)
- IdeaMaker (www.raise3d.com)

Most of these work with most of the popular 3D printers that are out there. When choosing slicing software, do make sure that your 3D printer can work with the slicer you're considering.

As for a recommendation, start with Cura, Simplicity 3D, Slic3r, or KISSlicer. Free versions and easy to learn and use.

You'll also need to consider if you want a 3D printer that can print multiple materials simultaneously. Basic versions of some slicing software will only work with one material. You'll have to upgrade (and probably pay for it) to work with multiple materials.

\\\ SELECTING A /// 3D PRINTER

There are 2 dominant types of 3D printers.

1. ***Fused deposition modeling (FDM)***, in which a spool of solid filament is heated and printed by a heated printer head called an extruder. FDM are the most common type of 3D printer. Because FDM 3D printers work with material called filament, they are often referred to as filament 3D printers in addition to FDM 3D printers.
2. ***Digital Light Processing (DLP)***, in which liquid material called resin is heated and then printed in similar fashion to FDM 3D printing. Because DLP 3D printers work with material called resin, they are often referred to as resin 3D printers in addition to DLP 3D printers. You may also see them referred to by their more technical names such as stereolithography, optical fabrication, and/or photo-solidification 3D printers. Again, the key difference between these types of 3D printers and FDM 3D printers is that FDM 3D printers use filament while DLP printers use resin as the material.

As you can see in the table below, there are definitely differences between filament and resin 3D printers. Resin 3D printers, overall, are on the higher price side, including the cost of the printer, cost of the material, and cost of maintenance and upkeep. However, resin 3D printers yield a much higher quality finished product. Because the resin is already in liquid form, resin 3D printers also tend to be faster than filament 3D printers.

	FDM or Filament 3D Printer	**DLP or Resin 3D Printer**
Printer Cost	Lower	Higher
Material Cost	Lower	Higher
Maintenance Cost	Lower	Higher
Print Speed	Slower	Faster
Quality of Finished Product	Lower	Higher

Most professionals will use both. They will use a filament 3D printer to do prototyping and then a resin 3D printer for the final quality print. If you're just starting out, buy a filament 3D printer first, a nice inexpensive option to get started.

Choosing a 3D printer is similar to choosing a regular paper printer… what capabilities do you need, how fast do you want to print, in what sizes do you want to print, do you need single or multiple colors, what's your budget? These are all important questions. In the 3D printing realm, they're a bit more complicated than a regular paper printer.

So, let's try to break this down as simply as possible, focusing on personal 3D printing needs.

- Price: They range in price from a few hundred dollars to a few thousand dollars. As we noted before, resin 3D printers are definitely more expensive than filament printers.
- Common/Exotic material question: Basic 3D printers, especially the inexpensive ones, work with only basic materials like the common ones previously listed. If you want to work with exotic materials, you'll need a more "advanced" 3D printer and one that costs more.
- Multiple material question: Basic 3D printers will only print from one material at a time. If you want to embed (i.e., print) magnets into your printed products because you're making wild and crazy refrigerator magnets, you'll need a more expensive printer that can simultaneously print from multiple filament spools or resin tanks. Some 3D printers can print from more than 2 filament spools or resin tanks; those are obviously more expensive. This also includes multiple colors of the same material.
- Build area: ***Build area*** addresses the maximum size of the object you can 3D print. Smaller, more inexpensive 3D printers have smaller build areas than larger and more expensive 3D printers that have large build areas. At the time we wrote this, typical sizes for build areas included: width 3 to 12 inches, depth of 3 to 8 inches, and height of 5 to 10 inches. The more width, depth, and height you want, the more expensive the printer.

\\\ 3D PRINTING IMPACT /// ON SUSTAINABILITY

Reduce-Reuse-Recycle… you gotta love it. Many of our big hairy audacious problems such as fast fashion, food waste, and the use of fossil fuels are antithetical to the notions of Reduce-Reuse-Recycle. And it takes innovation and clever thinking (and perhaps new technologies) to address those great big problems.

Think of Reduce-Reuse-Recycle as the big Rs. They're big because of their importance and because we typically address them at the macro level. Consider recyclable plastic soda and water bottles. They appear in mass quantities on grocery store shelves. We consume their contents in mass quantities. Then, hopefully, we throw them into blue/green recycle bins, and the recycle bins are hauled off by trucks to a recycle facility, where they are repurposed for use again. Rinse and repeat.

That's good and we have no problem with it. But what if you could reduce-reuse-recycle at the micro level, in your home? 3D reprinting offers us that opportunity.

3D reprinting uses existing items to recycle filament/resin and uses that recycled filament/resin to 3D print new versions of the same items (or entirely new items). Let's illustrate with a simple example, toothbrushes.

Someday, we'll 3D print our own toothbrushes. Really. Visit a website such as Thingiverse (www.thingiverse.com), find a toothbrush you like, download the STL file of it, and print all the toothbrushes you want. The cost? Maybe you'll pay $.99 for the download, although most likely the download will be free. When you need a new toothbrush, the material will probably cost less than 10 cents. You can have a new toothbrush every month of the year for about a dollar.

But what to do with the old toothbrushes? The solution is 3D reprinting, and it's already possible.

As an example, the ReDeTech ProtoCycler V3 allows you to recycle your old 3D-printed products and use the recycled filament of those to make new products. So, when you're done with your current toothbrush, use the filament in that toothbrush to 3D reprint a new toothbrush. You don't have to recycle your old toothbrush at the macro big "R" level. You simply do it in your home, the small "r"" level.

This can (and most likely will) happen with just about anything you care to name. Kitchen utensils - like spatulas, whisks, and stirring spoons - get old, worn out, or break. Throw them in your recycler and make new utensils from them. Break a shovel, make a shovel. Your favorite shirt gets old and worn out, make a new shirt from it.

What about one pair of shoes for life? When a child is born, parents can have a 3D-printed pair of shoes custom made for the child. Of course, the shoes won't last long because the child grows. When the shoes no longer fit, return the shoes, pick out a new style and size, and the company will use the returned shoes to 3D reprint a new pair of shoes. At the age of 18, when the parents order a new pair of shoes (and return the old ones), the child has a pair of shoes that are made from the exact material of the shoes he/she/they had at birth. What a beautiful circle-of-life story, and what a powerful example of small "r," reduce-reuse-recycle.

Think about candles. When a candle burns out, you throw away the remaining wax. Why not use the wax of the old candle to make a new candle? All you have to do is use the wax of the old candle to 3D reprint a new candle. Add the wick, which you can easily do with a 3D printer that works with multiple filaments/resin, and you have a new candle. You'll be able to easily change the scent and color of the candle. You can even reuse the candle holder.

BTW, wax already exists as an exotic material for 3D printing.

To truly understand the significance of 3D reprinting, think about interstellar space travel. (It's closer than you think.) When something breaks or wears out on a spaceship, you're not going to go out to the shed or supply room and find a replacement. You can't carry a backup of everything on a spaceship. Instead, you'll use 3D reprinting to use the filament of the old/broken item to make a new item.

\\\ THE END GAME: /// WHAT THE FUTURE MIGHT HOLD

3D Printing Everything
It's hard to imagine what we won't be 3D printing in the future. Perhaps big, huge pieces of furniture, but, then again, we're already 3D printing entire homes. Why not furniture?

DIY Meets PIY
Do-it-yourself meets 3D printing and becomes print-it-yourself. 3D printers will become so easy to use that you won't need to hire someone to design and print your fill-in-the-blank; you'll just do print-it-yourself.

Print-Recycle-Reprint
According to the EPA, the U.S. recycled 46 million tons of paper in 2018. [22] Let's recycle it at home and not just produce more paper, but produce paper that is automatically fed into a regular printer when we need to print on paper. Print-on-demand, paper-on-demand.

3D Printing Fake Fingernails or Directly on Fingernails
It's going to happen. The 3D printing of fake fingernails already has some limited success. 3D printing on the human body has as well. Just a matter of time before a 3D printer applies your fingernail polish.

3D Printing Makeup on Your Face
See above and substitute makeup for fingernail polish.

Edible Packaging
You gotta love this one. Think of the amount of trash and recycle we would eliminate if food came in 3D-printed edible packaging. Package of spaghetti noodles… just drop the whole thing - the noodles and the packaging - in hot water. Drink a bottle of soda, eat the bottle the soda came in. Eat a candy bar without ever taking off the wrapper. (Don't believe us… send us a notable quotable.)

CHAPTER 6

Autonomous Vehicles

The Death of the Driver's License

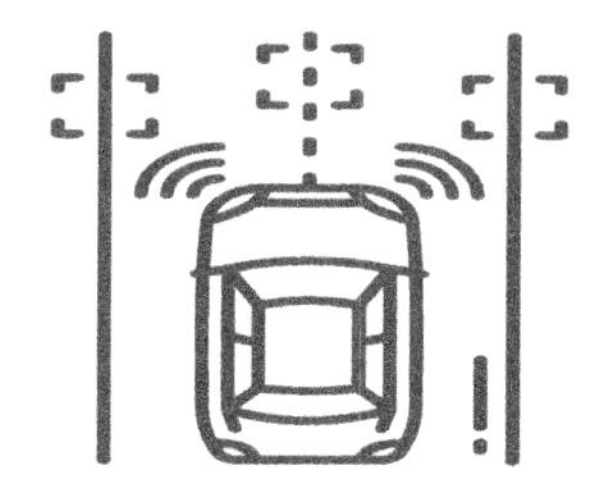

Time for you to answer another question. According to the National Highway Traffic Safety Administration (NHTSA), what percent of automobile accidents in the U.S. are caused by human error: [1]

A. 54%
B. 64%
C. 74%
D. 84%
E. 94%
F. 98%

May have fooled you here. The answer is E, 94% of all automobile accidents in the U.S. are caused by human error. The "errors" include things like excessive speeding, distraction (put down those phones), incorrect assumptions about what other drivers are going to do, tailgating, and not checking traffic lights before pulling out or changing lanes. [2]

- In 2022, 42,795 people in the U.S. died from automobile accidents. [3]
- Over 6 million automobile accidents happen each year in the U.S. [4]
- According to the NHTSA, 13,384 people died in drunk-driving crashes in the U.S. in 2021. [5] That means one person in the U.S. was a drunk-driving fatality every 39 minutes in 2021.

Grim statistics.

As one of the 7 core 4th industrial revolution technologies, we hope and believe that autonomous vehicles can reduce automobile accidents and deaths. Additionally, autonomous vehicles will enable significant innovation and industry disruption.

An ***autonomous vehicle*** is a vehicle that can guide itself without human conduction, using sensors, software, and a variety of driver-assistance technologies to navigate the vehicle. Driver-assistance technologies include things like adaptive cruise control, remote parking, and lane-departure warnings.

When we think about autonomous vehicles, we most commonly do so in terms of personal automobiles and cars. But *vehicle* can include long-haul trucks, airplanes, motorcycles, and any other type of vehicle that moves people, cargo, and the like.

Here, we'll be talking with you about autonomous vehicles mainly within the context of personal automobiles and cars. You can easily extrapolate to other types of vehicles.

\\\ THE RANGE OF AUTONOMY: /// FROM NOTHING TO EVERYTHING

The range of autonomy goes from none (completely manual) to completely autonomous, although we are not yet at the fully autonomous level.

According to the Society of Automotive Engineers there are 6 levels of driver-assistance technologies (i.e., autonomous vehicles). (See Figure 6.1) [6]

- Level 0 - no automation at all; the driver performs all driving and driving-related functions. At this point, you're getting no support, or perhaps very limited support only in the form of warnings like lane departure and blind spots.
- Level 1 - some support for (1) steering (e.g., lane centering) or (2) braking and accelerating (e.g., adaptive cruise control).
- Level 2 - some support for (1) steering (e.g., lane centering) and (2) braking and accelerating (adaptive cruise control) at the same time.
- Level 3 - the car can drive itself under certain limited conditions. The car may require that you take control of driving. (This level would include remote parking.)
- Level 4 - same as #3 (driving under limited conditions), except that the car will not require that you take control of driving.
- Level 5 - the car can drive itself everywhere under all conditions.

Figure 6.1 Levels of Autonomy in Autonomous Vehicles

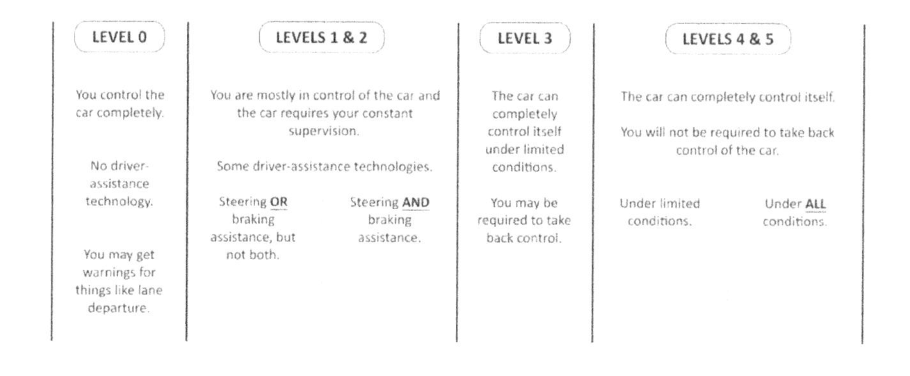

It's important to understand that <u>you are the driver completely</u> in the first 3 levels (0 through 2). At no time will your car take complete control of all driving functions. You must be in the driver's seat at all times, ready to take control of your car at any time. In level #5, alternatively, there may not even be things like a steering wheel and pedals present in the car. Won't that be wild to see? Levels 3 and 4 are obviously somewhere in the middle.

\\\ MOST COMMON DRIVER-
/// ASSISTANCE TECHNOLOGIES

Driver-assistance technologies control specific aspects of your car's operation, mostly in steering, braking, etc. Technically, warnings for things like lane departure and driver fatigue are not driver-assistance technologies. They warn you but they do not take corrective actions themselves. However, you will see these types of warnings listed under driver-assistance technologies. (The list below includes them.)

Driver-assistance technologies include:

- Lane departure warning - usually a beeping sound when your car notices you are "drifting." Some vehicles also provide haptic feedback by making the steering wheel vibrate.
- Driver fatigue warning - usually a beeping sound or haptic feedback when your car notices that you seem to be steering in a less-than-optimal fashion (e.g., weaving within your lane).
- Blind spot warning - usually a beeping sound or light flashing when you're moving in the direction of an object, or an object is moving toward you from a blind spot direction.
- Automatic braking (collision avoidance) - when objects suddenly appear in front of the vehicle.
- Adaptive cruise control - set your speed and the distance you want to stay behind cars in front of you, measured in cat lengths. Your car will slow down when it gets within a certain distance of the car in front of it.
- Parallel parking - this certainly seems to be a lost art. No explanation necessary.
- Weather-related assistance - controlling windshield wipers based on moisture, moving between dim and bright lights based on fog/rain/snow, shifting into or out of all-wheel or 4-wheel drive based on road conditions.
- Remote parking - while not even in your car, your car navigates into and out of tight parking spaces.
- Hands-free navigation/steering - no explanation necessary.

"Limited Conditions"

The term *limited conditions* applies to levels 3 and 4. In those levels, your car can drive itself - including steering, accelerating, and braking. But your car will only be able to engage all those driver-assistance technologies if certain conditions are met.

Limited conditions already do and may in the future include:

- Highway driving only (not on busy city streets)
- (Good) weather
- (Minimal or no) traffic congestion
- (No) construction areas in which speed is limited
- Not in school zones when reduced speed is enforced

Those are the major ones, and they make sense. A long stretch of highway in good weather with little other traffic and no construction is optimal (right now) for an autonomous vehicle to fully take control of steering, accelerating, and braking.

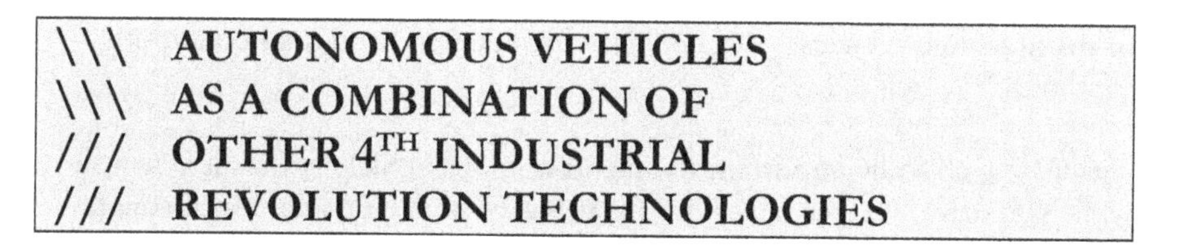

As a function of the other 4th industrial revolution technologies, autonomous vehicles look like this:

AV = AI + ST + CT + IoT (+Ba)

Autonomous vehicles, as a core technology in the 4th industrial revolution, are a combination of artificial intelligence (AI), sensing technologies (ST), communications technologies (CT), and IoT. We've included battery (Ba), as we believe all vehicles in the future will be fully electric.

Artificial Intelligence in Autonomous Vehicles

Artificial intelligence (AI) is key for autonomous vehicles. Autonomous vehicles must be intelligent and continue to learn even more as you drive (or as the car drives itself).

- Preloaded Intelligence - in the section on AI, we talked about how AI learns through training. When you buy an autonomous vehicle, it will come to you having driven perhaps millions of miles in a simulator. To be sure, the car itself (tires, engine, etc.) hasn't traveled millions of miles, rather the AI in the car has driven and learned from millions of miles. So, it will already know how to recognize and change lanes, how to accelerate and brake based on the traffic around it, recognize and interpret traffic signs, what to do at an intersection based on the lights and surrounding traffic, how to respond to the presence of pedestrians, cyclists, and so on.

- Continual Learning - As you use your autonomous vehicle (or allow it to drive itself), it will continue to learn. For example, it will learn more about your neighborhood, your garage for parking, and the peak and lull times for traffic getting in and out of your neighborhood.
- Collaborative Decision Making - Autonomous vehicles will communicate with each other and decide what's best for the collective of vehicles on the road and in proximity.

Sensing Technologies in Autonomous Vehicles

Just as you respond to hearing, seeing, and feeling (omitting smelling for obvious reasons) while driving, your autonomous vehicle will do the same using a variety of hearing, seeing, and feeling technologies. (We strongly encourage you to read Chapter 8 on Sensing Technologies.)

"Hearing" can include sirens of emergency vehicles and the backing-up beep of maintenance vehicles. "Feeling" can include wind and road conditions (potholes, ice, rain, etc.).

"Seeing" is critically important to an autonomous vehicle, just as it is to you while you drive. To see, autonomous vehicles use a combination of sonar, radar, and Lidar. Autonomous vehicles also use a variety of cameras - including 360 cameras - for seeing.

Sonar, Radar, and Lidar in Autonomous Vehicles

All three of these technologies are used in various forms by autonomous vehicles. Depending on the manufacturer, you will see a different level of focus for each technology. Elon Musk, for example, is a stronger advocate of sonar and radar as opposed to Lidar.

- Waymo (Google/Alphabet subsidiary) - uses Lidar as its main seeing technology. It also uses radar to help detect and interpret objects in rough weather conditions (fog, snow, etc.).
- Aurora Technologies (previously Uber's self-driving car initiative) - uses primarily both radar and Lidar.
- Tesla - uses primarily sonar and radar. Most Tesla vehicles have 12 ultrasonic sensor units, coupled with radar capabilities to detect objects up to 500 feet away.
- Rivian - has 12 exterior cameras, 12 ultrasonic sensors providing 360-degree close-range coverage, and 4 corner radars and 1 forward radar.

Regardless, it takes multiple seeing technologies for autonomous vehicles to be effective in driving without human conduction. Radar, for example, can detect objects up to 60 miles away. That's helpful for changing routes based on anticipated traffic congestion and accidents. But Lidar has a much higher degree of accuracy for detecting and identifying objects that are closer. That's helpful for determining if a stationary object is a signpost, a pedestrian, or a tree.

Communications Technologies in Autonomous Vehicles

When most people think about autonomous vehicles, they think about their own car driving itself around. That's a very narrow view; it may be true right now, but it won't be in the future.

As we get more autonomous vehicles on the road, they will communicate with each other to optimize driving. Examples:

- Once a traffic light turns green, all waiting vehicles will simultaneously move in tandem through the intersection. No delay because people aren't paying attention.
- When approaching a construction area with reduced lanes, all vehicles will communicate to determine the optimal way to combine lanes while maintaining speed. (Won't that be nice?)
- If your vehicle is ahead of mine, my vehicle will see what your vehicle sees, for example, a pothole.

Vehicle-to-vehicle communication is so important for autonomous vehicles. It will certainly provide for the optimization of driving in terms of time, speed, and other forms of efficiency. But it will also save lives. Right now, when a driver suddenly slams on the brakes, there is the potential for being read-ended because of the time it takes the driver behind to notice and respond. In the future, an autonomous vehicle will constantly inform other vehicles of what it is doing

IoT in Autonomous Vehicles

Just about everything we've discussed for sensing and communications technologies will be in the form of sensors and implemented via IoT. Cameras, Lidar, Bluetooth, WiFi, and many other sensing and communications technologies will be *a network of Internet-connected objects that collect, process, and exchange data* (our definition of IoT from Chapter 1).

\\\ LEADING COMPANIES IN /// AUTONOMOUS VEHICLES

As you can imagine, all the long-established car companies are working in the autonomous vehicle space… Toyota, Volkswagen, Daimler, Ford, Honda, BMS, GM, Mercedes-Benz… everyone. There are also many new players. Some are well-known like Tesla and Waymo, while a few names may be new to you.

The country of China is an interesting player, of sorts, in this market. While most other countries are researching and operating internationally in the autonomous vehicle space, China isn't (yet, although it is in the early stages of entering the European market). China-based companies include Pony.ai, AutoX, WeRide, Baidu, and DidiChuxing.

Consumer Autonomous Vehicles

The leading companies in the consumer autonomous vehicle space include:

- Tesla
- Apple
- Kia-Hyundai
- Ford
- Audi
- Huawei
- Rivian

Robotaxi Autonomous Vehicles

Robotaxis are exactly what the name implies, autonomous vehicle taxicabs. Waymo is the most well-known in this segment. Others include:

- Cruise (joint venture of Honda, GM, and a few others)
- Argo (Ford and Volkswagen)
- Motional (Hyundai)
- Zoox (Amazon)
- Aurora (previously Uber's self-driving initiative)

Long-Haul Truck Autonomous Vehicles

This space is interesting in that the transportation model being pursued by many companies is quite different from what most people think about. Those companies envision autonomous long-haul trucks only operating on the highways. Let's consider the Denver to Salt Lake City run.

There would be a transportation hub just outside of each of these cities. Human-driven trucks originating in Denver, for example, would drive a trailer to the transportation hub just west of Denver on Interstate 70. The trailer would then be connected to an autonomous long-haul truck which would drive it to the hub just east of Salt Lake City on Interstate 70. There, the trailer would be connected to a human-driven truck that would take it into the interior of Salt Lake City. In this way, the autonomous driving would occur only on highways.

Companies in this space include Kodiak, Embark, TuSimple, Waymo, and Aurora. The latter 2 are working in both this space and the robotaxi space.

\\\ THE DOWNSIDES TO AUTONOMOUS /// VEHICLES: IT CAN'T ALL BE ROSES

Every coin has two sides. Every rose has its thorns. Every night has its dawn. Say it however you want, but there are always goods and bads to everything. Autonomous vehicles are the same.

- Driver Job Loss - most notably among professional drivers… taxicab drivers, long-haul truck drivers, delivery drivers, and so on.
- Unintended Consequence #1 - because autonomous vehicles will reduce accidents, they will have a (seemingly) adverse impact on the automobile insurance industry, auto body shops, the tow truck industry, and so on.
- Unintended Consequence #2 - the majority of organ donations come from victims of automobile accidents. The wait time and list for organ donations are already long. They will get even longer with fewer organ donations coming from victims of automobile accidents. (Perhaps bioprinting can help. See Chapter 5 on 3D Printing, specifically bioprinting.)

There are undoubtedly many other "downsides" to autonomous vehicles. We have to carefully weigh all the advantages and disadvantages and make decisions in the best interest of society as a whole.

\\\ THE DECOUPLING OF AUTONOMOUS /// VEHICLES AND THE DRIVER

Once we get to fully autonomous vehicles (Level 5), everything will change.

You Can Do Anything
When you're in your car, you'll be able to do just about anything you want (within the law, of course). Automobile manufacturers are already rethinking the interior design of a car. Could it have a bed so you can nap on the way to work? Will the notion of a front seat and back seat go away, in favor of 4 chairs that face into the middle with a worktable in the center?

Do whatever you want… it will be like having a personal chauffeur.

Your Car Will Do Anything and Everything Without You
This will be interesting to see how it plays out. Certainly in your lifetime, your autonomous vehicle will drive you to work and then drive off to run errands, be an Uber driver, and go back to your house and take the kids to school. Your autonomous vehicle won't need you or anyone else as a passenger to go places and do things.

There will obviously be much debate and legislation around this issue.

\\\ THE END GAME: /// WHAT THE FUTURE MIGHT HOLD

The Death of the Driver's License
Someday, autonomous vehicles will not have a steering wheel, pedals, or any of the other gadgetry that is necessary for you to drive the vehicle.

When that happens, what's the point of having a driver's license?

The Death of Stop Signs, Traffic Lights, and Lanes
It's going to be so much fun to see what happens in this space. Think 20 years from now. We truly believe that all vehicles rolling off the manufacturing floor will be completely autonomous (yes, Level #5). In fact, 20 years from now you may have to get a special permit to drive a non-autonomous vehicle on the road. In that future, the "certain conditions" clause will apply to non-autonomous vehicles, not the reverse as it is now.

At that point, things like stop signs, traffic lights, and lanes will be non-existent. Just think about this with us for a moment. Why do we have stop signs at an intersection? Simple, we need them so drivers will know who has the right of way and be able to navigate without causing harm. In the future, all autonomous vehicles will approach an intersection and "negotiate" with all the other autonomous vehicles concerning the ordering of proceeding through the intersection. There won't be a physical sign telling your autonomous vehicle to stop. Using all the sensing and communications technologies, your autonomous vehicle will know that it can move on through the intersection because no other traffic is approaching the intersection.

In the same fashion, lanes won't be necessary. We've created lanes so people know how to position their cars. In the future, our autonomous vehicles will decide that. From east to west, for example, the collective of autonomous vehicles may decide to have 5 "lanes" of traffic going east in the morning and only 3 "lanes" of traffic going east in the afternoon. And that may change from day to day.

How many times and how much time have you wasted sitting at an intersection with a red light while no cars were traveling perpendicular to you? That won't happen in the future.

The Death of Rear-View Mirrors

We need them so we can see behind our own car while changing lanes, parking, etc. Autonomous vehicles will have cameras.

The Death of Windshield Wipers

We need them so we can see out the front windshield. Autonomous vehicles won't need to see through the front windshield. Makes you wonder… do we really even need glass we can see through? Your autonomous vehicle won't.

The Death of Headlights and Turn Signals

We think you're beginning to get the idea. Today's vehicles have lots of stuff to help us drive. Autonomous vehicles won't need things like headlights and turn signals. Think about the typical car today. What <u>won't</u> be needed in the future?

Autonomous Vehicles Impact on Rideshare

So much disruption is going to occur in the rideshare space over the next several years. The likes of the taxicab industry, Uber, and Lyft will undergo dramatic changes as vehicles no longer require a driver.

Moreover, we'll begin to see "share" more than "own." Think about this possibility. You're considering a number of different apartment complexes to move into. You have to think about location, price, size, and amenities. Those "amenities" may very well include the size of the apartment complex's autonomous vehicle fleet, how many free minutes you get each month included in your rent payment, and how many times each month you can use an autonomous vehicle during peak times.

HOAs (homeowners' associations) may offer this amenity for neighborhoods. Large retirement centers will undoubtedly do the same.

Mobile Shopping
Imagine a future with mobile shopping vehicles traversing through your neighborhoods. During winter months, the vehicles will travel around with an inventory of snow shovels, gloves, ice scrapers, and salt. You'll use an app to "hail" such a vehicle to your location. Once it arrives, you go in, find what you need, and pay using facial recognition software.

This will become a profitable business model because the vehicle won't have a driver. It can run 24x7, needing only to "take a break" to replenish inventory and get a battery charge.

Your Autonomous Vehicle Will Pay for the Toll Road
Think combining autonomous vehicles with other 4th industrial revolution technologies, like cryptocurrency. You'll travel down a toll road and your autonomous vehicle will use cryptocurrency to pay for the toll. Guess what, your autonomous vehicle will be connected to your bank account(s).

Your Autonomous Vehicle Will Pay for Parking
See above and substitute "parking" for toll road.

Your Autonomous Vehicle Will Pay for Car Washes
See above and substitute… well, you know what we mean.

Disruptions to Supply Chain Management and Delivery Activities
Coupled with drones (our next chapter), autonomous vehicles will most certainly redefine supply chain management and delivery activities.

Delivery vehicles will perform their tasks 24x7, like mobile shopping vehicles we just discussed. UPS, FedEx, and Amazon are already exploring delivery vehicles equipped with drones. The delivery vehicle itself may never stop, instead just drive through a neighborhood with drones launching from the top, dropping packages at homes, and then flying to the location of the delivery vehicle to get the next payload.

When this happens (and we believe it will), supply chain management goes from a 2-dimensional optimization problem to a 3-dimensional one. More on this in the next chapter on drones.

Out with the New, In with the Old

Someday, we won't call them autonomous vehicles. We'll simply call them *cars*. (By the way, the word *car* comes from the Celtic word *carrus*, for 'cart' or 'wagon.' It was in fact first used to describe a horse-drawn carriage.)

CHAPTER 7

Drones

Male Bee, Bagpipe, Bladder Fiddle, or Flying Vehicle?

Time for another question. How would you define a drone?

A. Flying vehicle
B. Remote controlled aircraft
C. Unmanned aircraft
D. Male bee from an unfertilized egg
E. Musical note or cord
F. Bagpipe
G. Bladder fiddle
H. Type of minimalistic musical style

Trick question, of sorts, as all the answers are correct. Drones, as we'll discuss them here, run the full gambit across many perspectives including aerial to underwater, commercial to personal, remote-controlled to autonomous, and so on. That makes it difficult to provide any sort of succinct encompassing definition. Let's just say ***drones*** are flying vehicles without needing a pilot in the vehicle. (That still isn't completely correct as remote-controlled underwater vehicles are often called underwater drones.) They may be remote-controlled or completely autonomous. You may or may not need a license to fly one. Drones may or may not have people in them. Drones, what a term. (And what in the world is a bladder fiddle?)

Quick Lesson in Aviation

Before we launch (yes, pun intended) into our discussion of drones, let's have a quick aviation lesson to get a few terms in place.

- Fixed wing aircraft generate forward thrust. Commercial airline flights you take are in a fixed wing aircraft. The speed of the aircraft, the flow of the air, and the shape and tilt of the wings generate lift, causing flight.
- Rotary wing aircraft generate vertical thrust. Just think about a typical helicopter. Enough said.
- Single engine aircraft have a single engine for providing thrust. For fixed wing aircraft, there is a single engine that propels the aircraft forward. For a rotary wing aircraft, there is a single engine that spins a single rotary blade.
- Multi-engine aircraft have multiple engines for providing thrust (duh). For fixed wing aircraft, there are multiple engines that propel the aircraft forward. For a rotary wing aircraft, there are multiple engines that each spin a different rotary blade. The latter type of aircraft is often called a multicopter.
- ***Unmanned aerial vehicle (UAV)*** - aircraft without any pilot, crew, or passengers aboard but is remotely piloted. Think personal drone. You may also see r*emotely piloted aircraft (RPA)* used interchangeably with UAV.
- ***Autonomous drone*** - aircraft with the necessary software, sensors, and artificial intelligence to fly itself without human conduction.

\\\ CONSUMER VERSUS /// COMMERCIAL DRONES

There are significant differences between consumer (personal) and commercial drones.

Consumer Drones

Consumer drones are for your own use and enjoyment. You use a set of remote controls to fly the drone. Most consumer drones today are quadcopters because they have 4 rotary blades. You can take photos around your house. You can survey land you own. You can even engage in business activities such as using a camera on a personal drone to take aerial photos of a home that you want to sell. Popular consumer drones include: [1]

- DJI Mini 3 Pro Camera Drone
- DJI Air 2S Drone Fly More Combo
- Ruko U11Pro Foldable Drone
- DJI Mavic Mini 2 Fly More Combo Drone
- Holy Stone GPS RC Drone
- Ruko F11Pro Drone with Camera
- DJI Mavic Air 2
- Potensic ATOM SE GPS Drone With 4k Camera
- DEERC DE22 GPS Drone
- DJO Mini 2 SE Drone
- DJI Mavic 3 Cine Combo Drone
- Holy Stone HS720 GPS Drone
- DJI FPV Combo Drone

When purchasing a consumer drone, you'll want to think about (besides price):

- Flight time per battery charge (ranges from 15 minutes to about an hour)
- Speed (somewhere between 45 and 70 mph, typically)
- Smart home return (just press a button on your remote controls and the drone will return to you; some drones will do this automatically when the battery charge gets low)
- Environment awareness ("seeing" obstacles and the ground to avoid crashes)
- Noise (some make little noise, others are very loud)
- Camera (some have built-in camera capability, others have attachments for cameras)

Consumer Drone FAA Regulations

The FAA regulations regarding consumer drones are changing all the time, as we begin to see more widespread use of drones. The FAA's Small UAS Rule (Part 107) addresses the regulations and requirements. These include: (1) obtaining a Remote Pilot Certificate, (2) registering your drone with the FAA, and (3) meeting several basic requirements such as age (you must be 16 or older) and being able to read, speak, write, and understand English. These requirements relate to drones that weigh 8.8 ounces or more. Most consumer drones do. (A drone that weighs less than 8.8 ounces is considered to be just a toy.)

Beyond those initial requirements, you must follow the rules of the air:

- Fly at or below 400 feet
- Keep the drone within sight
- Don't fly over restricted airspaces, near other aircraft, over groups of people, over stadiums or sporting events, or near emergency response areas such as road accidents and forest fires
- Finally, don't fly under the influence

Most cities, municipalities, and homeowners' associations have or are developing additional regulations regarding where and even when drones can be flown.

Commercial Drones

Commercial drones are obviously a significant step up from consumer drones across all characteristics of price, speed, battery life, payload, etc. And the range from "small" to "large" is significant. Smaller commercial drones may have a relatively short battery life (30 minutes) and a maximum payload of a couple of hundred pounds. On the other end of the spectrum are commercial drones used by the military. These can easily weigh tons (literally) and have large payload capacities.

Somewhere in between are the commercial drones that the likes of Amazon, UPS, FedEx, and many other commercial businesses are exploring for package delivery.

\\\ DELIVERY COMPANIES /// AND DRONES

The race is certainly on to use drones for deliveries. Drone package delivery helps minimize the use of road vehicles, is environmentally friendly because of the use of electricity, and can shorten delivery times because of direct flight capability instead of having to navigate winding roads, traffic congestion, intersections, and the like.

Consider these:

- Zipline is using drones to deliver blood and other medical supplies. [2]
- Kroger and Drone Express have partnered to deliver groceries to consumers. [3]
- UPS has received FAA authorization for a full drone fleet to fly as many delivery drones as it wants. [4]
- FedEx has received FAA authorization as well. [5]
- Amazon has PrimeAir, which will deliver packages up to 5 pounds in 30 minutes or less. [6]

Other Innovative Uses of Drones

When most of us think of drones, we do so in terms of delivering packages because that's what we most often see in the popular media. Package delivery is going to be big in the drone business, but there are and will be others.

- First responder, police and law enforcement, and public safety [7]
- Agriculture - for reducing the time to scout crops, field mapping, and spraying [8]
- Real estate - for surveying land, providing aerial footage of homes, etc. [9]

\\\ DRONES AS A \\\ COMBINATION OF OTHER /// 4TH INDUSTRIAL REVOLUTION /// TECHNOLOGIES

In formulaic terms, it looks like this.

D = AI + ST + CT + IoT (+Ba)

Like autonomous vehicles, drones are a combination of artificial intelligence (AI), sensing technologies (ST), communications technologies (CT), and IoT. We put batteries (Ba) in parenthetical marks because, while most (smaller) drones are electric-powered, some really large commercial drones use fossil fuels (gasoline, kerosene, and the like).

Artificial Intelligence in Drones

Today, artificial intelligence may or may not be present in a drone, but it certainly will be in the future. For a drone to be autonomous, it must have sophisticated AI, just like an autonomous vehicle must.

We are seeing some "intelligence" added to drones today like environment awareness to avoid collisions with the ground, telephone poles, other drones, and the like. That's most probably a form of reactive AI, which simply detects an object and moves through a series of rules to determine what action to take.

The future challenge of fully autonomous drones is two-fold.

The first is the *added-dimension challenge.* Autonomous vehicles work in only two dimensions, while autonomous drones will have to work in three. So, not only must autonomous drones be able to navigate in that third dimension, we'll also have to map the third dimension. Most mapping functions like Google Maps are great with respect to longitude and latitude, but now we'll also need the height of homes, buildings, utility poles, and even trees.

The second is the *circular challenge.* As we add more AI-based capabilities to support fully autonomous drones, we'll essentially be adding more weight to the drone. As the weight of the drone increases, you need more battery capacity. More battery capacity means more weight. More weight means you need more battery capacity. More battery capacity means… well, you get the idea.

Sensing Technologies in Drones

Again, just as autonomous vehicles use a variety of hearing, seeing, and feeling technologies, so do drones, especially as they become more autonomous.

Hearing, seeing, and feeling will be necessary for drones to detect and respond to the presence of other objects, both fixed and moving (like and including birds). Drones will also need seeing technologies like sonar, radar, and Lidar to determine a flight route and respond in real time to changing conditions. Drones will need sophisticated feeling technologies to detect and respond to changes in the wind and wind shear, both horizontally and vertically.

Some drones will most likely use smelling technologies as well, perhaps more so than autonomous vehicles. As a scouter, the National Forest Service uses drones to fly over remote areas to detect the presence of fires. Fire detection can occur through video and image recognition but also through "smelling" the presence of smoke (using sensors to intake air and measure for the presence of chemicals/gases that would signal a fire).

And, as we introduce more of these sensing capabilities, we'll have to address the circular challenge. As you add more hearing, seeing, feeling, and smelling technologies, you increase the weight of the drone. And we know where that takes us… round and round.

Communications Technologies in Drones

Even the most basic drones require communications, at a minimum from the remote controls you operate to the drone itself. More advanced communications technologies will be a necessity as well. If you're delivering a package to someone via drone for example, you'll need GPS capabilities to determine the location of that person.

And eventually, we'll need drone-to-drone communications. Much like autonomous vehicles will communicate with each other to make collective decisions, so will drones.

IoT in Drones

Just about everything we've discussed for hearing, seeing, feeling, smelling, and communications technologies will be in the form of sensors and implemented via IoT. Camera, Lidar, Bluetooth, 5G, and many other sensing and communications technologies will be *a network of Internet-connected objects that collect, process, and exchange data* (our definition of IoT from Chapter 1).

\\\ FLYING CARS – /// eVTOL AND AIR TAXIS

eVTOL (electric vertical take-off and landing) refers to the ability of an electric-powered aircraft to take off, hover, and land vertically. So, eVTOL aircraft do include personal and (most) commercial drones and can also include air taxis, basically the equivalent of a flying car. Consistent with the architecture of a drone, eVTOL aircraft require the use of multiple blades, the *multicopter* concept.

Flying cars might seem a bit far-fetched and almost sci-fi, but they will be a big part of your future, and they will arrive sooner than most people think. Companies like Toyota, Joby Aviation (which acquired Uber Elevate), Hyundai, Airbus, and Boeing are all racing to perfect the flying car.

Flying cars hold a lot of promise… reduced traffic congestion on the roads, much faster "as the crow flies" times from suburban housing to city-center workplaces (and back), reduction in carbon emissions because of the electric power, faster responder times to accidents, etc.

It's probably important to keep in mind that flying cars will first appear as taxis or rideshares. The initial cost and maintenance expenses will likely prohibit most people from owning a personal flying car. But, over time, that will change. Won't that be interesting to see.

Flying cars are going to be a fascinating development to watch. Below are a few initiatives:

- City Airbus NexGen by Airbus [10]
- SkyDrive, a Japanese startup backed by Toyota [11]
- Joby Aviation [12]
- Hyundai and Supernal flying taxi [13]
- Boeing [14]
- Pal V Liberty, Samson Switchblade, Aska A5, Klein Vision AirCar, Alef Model A, Doroni H1, and Maverick Flying Car [15]

\\\ THE END GAME: /// WHAT THE FUTURE MIGHT HOLD

Instant Pudding from the Air
It's only a matter of time before we instantly get our deliveries via drone. Everything from fast food to BBQ sauce to sunscreen delivery will find you from the air no matter where you are.

Autonomous Vehicles and Drone Delivery
It will probably be advantageous for the big delivery companies - Amazon and the like - to use delivery trucks combined with drone capability. Both the delivery trucks and the drones will be autonomous. The delivery trucks will move through a neighborhood and act as a mobile launch pad for the drones. Each drone will take a package from the truck to a destination and then fly back to the new location of the truck to get the next package. And several drones will be doing this simultaneously, really optimizing the time to deliver packages.

Indoor Drones
We're already beginning to see the emergence of indoor drones. For example, the Ring Always Home Cam is an indoor drone with camera capability. [16] Now, you don't need a dozen or so security cameras to see everywhere in your home.

Drone Pooper Scoopers, Leaf Blowers, Window Washers, Pothole Finders, House Painters, Gutter Cleaners, Tree Trimmers, and More
This is really going to be a fun space… the innovative use of drones we'll see for taking over personal, mundane, and un-fun tasks at home. We've already got drone pooper scoopers, what's next? Perhaps drones that can blow the leaves off your yard, wash your windows, paint the house (don't laugh, it will happen), trim your trees, put up and take down decorations, clean the gutters, change light bulbs. Will ladders go the way of the dodo bird?

A Different Way to Watch Sports
This one kind of doesn't need any sort of explanation. Unlimited viewing perspectives of sporting events through the use of drones. Just have to be sure and avoid the ball.

Smaller and Smaller
We do now have what are called nano or mini drones, about the size of the palm of your hand. Many include camera capability. Payload is almost nil, but that will change over time. Imagine drone-mounted speakers for your TV viewing experience. Real spatial audio.

Crowded Airways
Someday we think it will be common to look up in the sky and see perhaps upwards of a hundred or even a thousand drones flying around. It may seem unrealistic, but in the early 1900s people couldn't envision that someday there would be so many cars that we would need multiple lanes of traffic all going in the same direction. (Some people believed there would be so few cars that we wouldn't ever need traffic signs, much less traffic signals at an intersection.)

Enter the government. There will undoubtedly be significant legislation and laws regarding use of the airways by drones (and flying cars).

Pads and Parking
Just as public transportation (buses, subways, etc.) has necessitated the need for terminals and parking around the terminals, so too will flying taxis. We'll essentially need helo pads for take-off and landing. And we'll need places for people to park their cars. Of course, then again, maybe not. Perhaps your autonomous vehicle will drive you from your home to the flying taxi pad and drop you off, then return to your home. Hmmm.

CHAPTER 8

Energy Technologies

Harvesting and Storage

\\\ THE WORLD /// OF ENERGY

Energy… interesting phenomenon. If someone asked you to define energy, what would you say?

(Don't worry, most people find it challenging to provide a simple and short definition of energy.)

Technically, ***energy*** is the ability to do work, which makes about as much sense as doing an interpretive dance to the smell of the color nine. (Noodle on that for a while.)

So, let's explore energy, talk about what we're doing in this space, and how energy will play out in the 4th industrial revolution.

The Many Forms of Energy

Energy comes in many forms, including:

- Heat/Thermal
- Light/Radiant
- Motion/Kinetic
- Electrical
- Chemical
- Gravitational
- Magnetic
- Mechanical (many forms including pressure)
- Elastic… just to name a few

What's really cool about energy is that it's *not* consumed in the traditional sense of consumption. That is, one type of energy simply transforms into another type of energy.

- When you eat food, the food contains chemical energy which your body stores and eventually converts to kinetic energy when you move.
- When you rub your hands together, you are converting kinetic energy (motion) to thermal energy (heat). (You're also transferring kinetic energy from the motion of your hands to the movement of the air.)
- When you use a flashlight, the energy in the battery is stored in chemical form. The chemical energy is converted into electrical energy which powers the bulb. The bulb uses the electrical energy to create radiant energy (light).
- When you drop something, the gravitational energy that pulls it to the ground is converted into kinetic energy.

Of course, energy is stored in chemical form in things like coal and natural gas, which we often convert to other energy forms such as electricity.

There you have it. You probably know more about the basics of energy than 99% of the world's population. (Although, you might not want to put that on your LinkedIn profile.)

How We Measure Energy

Each type of energy has its own specific measure, such as watts and kilowatts used to measure electricity. As a single common measure, we convert energy-specific measurements to a British Thermal Unit . A ***British Thermal Unit***, or ***BTU***, is the amount of heat required to raise the temperature of one pound of water by one degree Fahrenheit. (One pound of water is just shy of 16 ounces, or 2 cups of water.) In Figure 8.1, you can see that the U.S. consumed 100.41 quadrillion BTUs (that's 100,410,000,000,000,000 BTUs) of all types of energy in 2022. [1] (We heat a lot of water.) Most of it, almost 80%, comes from petroleum, natural gas, and coal (fossil fuels).

Figure 8.1 **U.S. Energy Consumption, 2022**

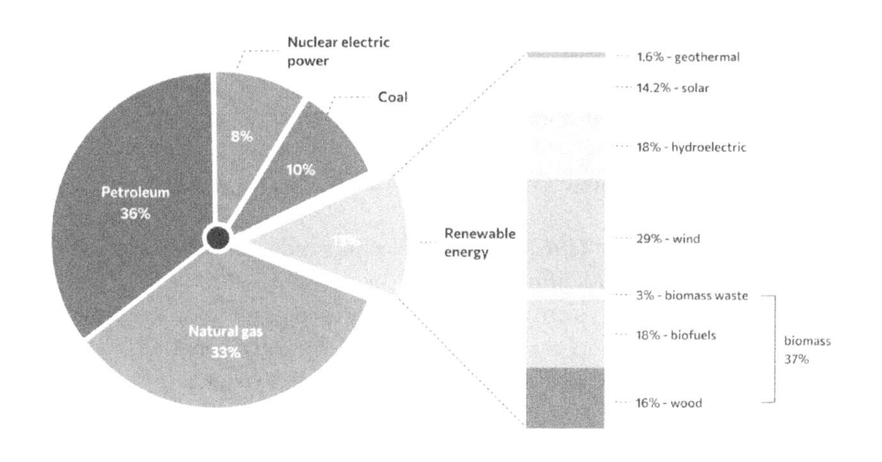

Renewable Energy and Its Importance

Renewable energy is energy that comes from naturally-replenishing and carbon-neutral sources, that include:

- Solar
- Wind
- Hydropower (water movement)
- Geothermal (heat from the earth's core)
- Biomass (wood and wood waste, municipal solid waste, landfill gas and biogas, ethanol, and biodiesel)
- Hydrogen
- Piezoelectricity

Take wind energy, as an example. If you have a wind turbine that captures the energy of the movement of wind today, that doesn't diminish the amount of wind there is tomorrow. The same is true of solar energy. Capturing the sunlight (radiant energy) today doesn't diminish the amount of radiant energy produced by the sun tomorrow.

Fossil fuels are not considered renewable energy. It can take millions, and sometimes hundreds of millions, of years to slowly convert decomposing plants and animals into fossil fuels.

And as you're probably aware, we want to get away from our reliance on fossil fuels because our use of those is the largest source of carbon dioxide emissions, a significant component of greenhouse gases.

The Impact of Greenhouse Gases on Global Warming & Climate Change

Greenhouse gas (***GHG*** or ***GhG***) is a gas that emits radiant energy causing the greenhouse effect. The ***greenhouse effect*** is the warming of the earth's surface from radiation in the earth's atmosphere. The primary greenhouse gases are water vapor, carbon dioxide, methane, nitrous oxide, and ozone. Of those, carbon dioxide is the most prevalent in causing global warming.

Global warming is the increase in the overall temperature of the earth's atmosphere, mainly occurring because of the greenhouse effect through the increase in carbon dioxide levels. Global warming and climate change are <u>not</u> the same. ***Climate change*** refers to long-term shifts in temperatures and weather patterns. So, climate change includes both global warming (an increase in temperature) and global cooling (a decrease in temperature).

The Personal Cost of Climate Change

The global impact of climate change translates directly into personal costs for you. [2]

- The U.S. Government has spent almost a half-trillion dollars since 2005 on climate change-related disasters. Your tax dollars paid for that.
- Average monthly electric bill: $104.92 in 2009 to $115 in 2019, that's an annual increase of $120 to your electric bill.
- 17.6% increase in beef prices from 2019 to 2021 because of drought, supply chain issues, and labor shortages.
- 2.5% increase in food-at-home prices from 2020 to 2021.
- Catastrophic weather events (wildfires, hurricanes, floods, etc.) were the cause of 39% of all insurance claims in 2020.
- Over a 2-year period of time, insurance companies increased insurance rates by 10% to 20% to address the impact of climate change.

Climate change is real, very real within the global context and in your wallet.

As we are in a period of global warming, our focus is on the increase in greenhouse gases, how they impact the greenhouse effect, and the subsequent rises in temperature.

Greenhouse gases, like carbon dioxide, are actually good; we don't want to get rid of them all. In fact, if we had no greenhouse gases, the earth's temperature would plummet to about 0^0 F (severe and catastrophic global cooling.)

Thus, there is a delicate balance we need to achieve. Greenhouse gases essentially keep the earth warm. But, if greenhouse gases become too much through things like excessive carbon dioxide emissions, then the earth's temperature rises, thus causing global warming.

\\\ ENERGY HARVESTING INSTEAD /// OF ENERGY MANUFACTURING

Our use of fossil fuels comes from *energy manufacturing.* When we think about energy manufacturing, we envision large processing plants - smoke stacks reaching to the sky and billowing smoke, the blackness of coal, the iron and steel of vast numbers of railroad tracks and cars, and so on. Those types of energy manufacturing plants are using fossil fuels as raw materials, transforming the chemical energy in them to another type of energy.

What we'd really like to do is ***energy harvesting***, capturing a renewable energy form and converting it into another energy form that is usable "on the spot." Think of solar panels you see on your neighbor's roof. They are capturing the energy from the sun (solar source for light or radiant energy) and converting it into electricity for immediate use in the home.

To harvest energy as opposed to manufacturing it, we need to focus more on:

- Solar power
- Wind power
- Hydropower
- Geothermal power
- Biomass
- Hydrogen
- Piezoelectricity

Solar Power

Time for another question: One second of the sun's total energy output would power the United States for how many years?

A. 900 years
B. 9,000 years
C. 90,000 years
D. 900,000 years
E. 9,000,000 years

The answer is E: 9,000,000 years. [3] Unbelievable when you think about it. If we could figure out a cost-effective and environmentally friendly way to harness the power of the sun, we would solve all our energy problems. End of story.

Formally, ***solar power*** is the conversion of sunlight (radiant energy) into electricity. We hope and believe that solar power will become the world's largest source of electricity by 2050. (We should work diligently to make that happen sooner.)

A lot of progress is being made in this area, and there are many ways to capture the radiant energy of the sun, with the two most common being:

- ***Photovoltaic system (PV system)*** uses solar panels to absorb and convert sunlight into electricity. Those solar panels on your neighbor's roof are a PV system.
- ***Concentrated solar power (CSP)*** uses mirrors and lenses to concentrate a large area of sunlight onto a receiver, which converts it to heat which drives a heat engine/steam turbine connected to an electrical power generator. CSPs are much larger in size than PV systems, so you won't find them on your neighbor's roof. These often occupy several acres of land. The largest CSP in the U.S. is Ivanpah, 3,500 acres in the California desert. It can generate 940,000 megawatt-hours of clean energy per year, preventing 500,000 metric tons of carbon dioxide emissions annually. [4]

Besides converting solar energy (i.e., radiant energy) into electricity, we're also seeing a number of applications using the thermal energy of the sun.

- Solar water heating, especially in the middle geographical latitudes (Ethiopia, as an example) [5]
- Heating, cooling, and ventilation in the home
- Desalination (making potable water from saline and brackish water) [6]
- Solar water disinfection (SODIS), exposing plastic bottles of water to sunlight. Millions of people in developing countries use this method for their daily drinking water. [7]

Mobile Solar (MoSo) - The Time Is Now

Our traditional view of solar power is large fixed-in-place solar panels generating energy for homes, cars, commercial buildings, streetlights, etc. The next evolution is for all of us to take advantage of mobile solar. ***Mobile solar (MoSo)*** is the use of small portable solar power units that you can take with you wherever you go.

MoSo units usually weigh a couple of pounds or less and enable you to recharge smaller electronics such as your phone and tablet. Throw a MoSo on your backpack while hiking, place it on a table while you're enjoying a day in the sun… whatever. MoSo units will save you a couple of dollars per year in electricity charges per device. That seems small. But if we all do it, we could collectively save billions of dollars per year in electricity costs.

Check out these options:

- Hiluckey Outdoor Portable Power Bank
- CONXWAN Solar Charger

- FOCHEW Solar Portable Charger Power Bank
- YPWA Portable Charger Power Bank
- Sunnybag Leaf (Mini and Pro versions)

Each cost less than $50.

Wind Power
We've taken advantage of wind power for thousands of years. Sailing ships in the Viking age, for example, captured the power of wind in their sails to propel the ship across the water (kinetic energy). Of course, now we're focusing on capturing wind power and converting it into electrical energy.

Wind power is the capturing of the kinetic energy of the wind to power wind turbines which provide mechanical power to electric generators which in turn generate electricity. Most often, we capture wind power using a ***wind farm***, a collection of wind turbines in close proximity that power numerous electric generators that are connected to an electric power transmission network. The power transmission network moves the electricity from the energy generation source (the wind farm) to homes, businesses, cities, and grid storage units.

The two dominant types of wind farms are onshore and offshore. Although we shouldn't have to say this… onshore wind farms are a collection of turbines on land, mostly in rural areas, while offshore wind farms have their turbines on water. Onshore wind farms are much less costly to build and maintain than offshore wind farms, but many people believe that onshore wind farms are leading to habitat loss (animals) and the industrialization of the countryside. Offshore wind farms, again while expensive to build and maintain, do have a steadier and stronger supply of wind. And, yes, many people do question the negative ecological impact offshore wind farms have on the ocean environment and marine life. (BTW, both types of wind farms have been criticized for the number of birds that die in wind turbine collisions each year.)

Wind turbines like the ones in the accompanying image have been the "standard" in looks for many years. These types of wind turbines are very, very large. Some reach almost 500 feet in the air and have blades that are almost 200 long. A single wind turbine of this size can power 1, 200 to 1,500 average-sized U.S. homes. But, because of their size, they have often been criticized for their lack of visual appeal.

A recent innovation is the *flower wind turbine*, which looks much like a tulip. The overall structure of a flower wind turbine typically does not exceed 10 feet. Their blades are about 6 feet in length and about 3 feet in width. Flower wind turbines produce much less energy than a traditional-looking wind turbine but have much more visual appeal. Many flower wind turbines have brightly colored blades, often with each blade a different color.

While on their own, flower wind turbines produce much less energy than a traditional wind turbine. However, you can put together numerous flower wind turbines to create a "bouquet effect." That is, a bouquet of 5 flower wind turbine will increase energy output by over 200%. The more dense the bouquet, the better the energy output.

Read more about flower wind turbines at https://www.flowerturbines.com/benefits.

Like solar, if we could figure out a cost-effective and environmentally friendly way to harness the power of the wind, we would solve all our energy problems. (Different solution to the same problem.)

Hydropower

Hydropower or ***water power*** is the capturing of the kinetic or gravitational energy of running or falling water to run a turbine or series of turbines which in turn generate electricity. This is achieved in a hydroelectric power plant. Hydroelectric power plants are often thought of as human-made dams that create a reservoir of water so that there is a consistent flow of water to the turbines. But they can also be what are called run-of-river, which doesn't use a dam to create a reservoir. Run-of-river hydroelectric power plants rely on a consistent flow of water to meet energy needs.

The most well-known hydro-electric power plant in the U.S. is the Hoover Dam, completed in the late 1930s, between the borders of Nevada and Arizona on the Colorado River. Annually, it can produce enough electricity to meet the needs of 1.3 million people. (Put the Hoover Dam on your bucket list. Take the long tour with a guide. Fascinating story and structure.)

Geothermal

Geothermal power takes on two forms, either (1) electricity generated from geothermal energy or (2) heat generated from geothermal energy. In either case, the naturally occurring thermal energy in the earth's crust is used.

Geothermal energy has actually been around for thousands of years, dating back to the Paleolithic times, and includes:

- Oldest known spa - 3rd century BC Qin Dynasty in China
- Public baths and underfloor heating - Bath, Somerset, England (1st century)
- Oldest geothermal district heating system - France, 15th century
- First known building to use geothermal energy for heating - Hot Lake Hotel in Union County, Oregon

In the early 21st century, we built the first successful geothermal power generator which used geothermal energy to generate electricity.

Biomass

According to the 2022 data in Figure 8.1, biomass accounted for 37% of all renewable energy in the U.S.. Renewable energy accounted for 13% of total U.S. energy use, so biomass accounted for roughly 5% of U.S. energy consumption. ***Biomass*** is the use of plant and animal material as fuel to generate electricity or heat. Biomass includes things like:

- Biomass waste - waste from forests, yards, and farms (bark, sawdust, wood chips, wood scraps, etc.)
- Biofuels - energy crops or crops grown specifically for energy production (willow trees, poplar trees, and elephant grass are examples); ethanol and biodiesel also fall into this category.
- Wood - you guessed it, the burning of wood for cooking, lighting, and heating. Wood was the primary source of energy in the U.S. up through the mid 19th century.

Hawaii - Last State In But First to Go Green [8]

The state of Hawaii has an abundance of solar, wind, and geothermal energy but not a single drop of extractable fossil fuels beneath its surface. All fossil-based fuels must be imported.

That reliance on imports has led Hawaii to focus its energy efforts and utilization on renewable energy.

- 32% of all its energy comes from renewables
- 1 in 3 people have PV solar systems for their homes
- 15% of new cars are electric
- The Kapolei Energy Storage Facility, about 20 miles west of Honolulu, has 158 storage sheds of lithium iron phosphate batteries that can store enough electricity to power 17% of the island of Oahu for 3 hours during peak load or 6 hours at half load.
- On the big island of Hawaii, approximately 30% of its energy comes from a geothermal plant that harvests heat from the Kilauea volcano.

Hydrogen

Not to be confused with hydropower, hydrogen fuel also holds potential as a renewable energy source. A ***hydrogen fuel*** cell produces electricity by combining hydrogen and oxygen atoms. A hydrogen fuel cell is 2 to 3 times more efficient than a traditional internal combustion engine running on gasoline. Even better, the only byproduct is water, so it falls into the category of *zero-emission.*

While there is much interest and research in hydrogen fuel cells, there is very limited commercialization in the electric vehicle space. As of early 2021, there were fewer than 50 hydrogen fueling stations in the U.S., with most of those being in California. [9]

By the way, Elon Musk referred to hydrogen fuel cells as "mind-bogglingly stupid." But a 2017 survey of 1,000 automobile executives yielded that hydrogen fuel cells ultimately would outperform battery-powered electric vehicles. [10]

So, an energy source and technology to watch.

Piezoelectricity
This has tremendous potential but is still in its early stages of development. ***Piezoelectricity*** is created when you apply pressure to specific types of materials (for example, certain types of crystals and ceramics). You can, for example, embed piezoelectric energy harvesting plates in a sidewalk. When you walk on the plates, the pressure of your steps creates electrical energy that can be harvested.

Applications of using piezoelectricity include: [11]

- Tokyo subway - as travelers walk on the plates, the captured electricity runs turnstiles, displays, and lights
- Sainsbury's - an English supermarket capturing electricity generated by shoppers
- Dance club - In Rotterdam, uses captured electricity to run the dance floor light shows
- Sidewalk - In Toulouse, France pedestrians create electricity that is used by the city to power light poles

Some people are even proposing that we embed piezoelectric energy harvesting plates in roadways to generate electricity from and for the cars driving over them. California already has a 60-meter stretch of roadway near Fresno that uses this concept. [12]

The Paris Agreement and Climate Change

The ***Paris Agreement*** (also called ***Paris Accords***, ***Paris Climate Agreement***, or ***Paris Climate Accords***) is an international treaty on climate change adopted in 2015 by 197 countries. The main goal of the Paris Agreement is to substantially reduce greenhouse gas emissions in an effort to limit global temperature increases in the 21st century to no more than 2 degrees Celsius above pre-industrial levels. The big hairy audacious goal is to limit global temperature increases to 1.5 degrees Celsius.

The Paris Agreement is a commitment by all countries to reduce pollution, with a big focus on carbon dioxide reduction. It also includes a commitment by developed nations to assist less-developed countries in their efforts to address climate change, through financial resources and the sharing of technologies. Finally, the Paris Agreement provides for a transparent framework for monitoring and reporting the efforts of each country to address climate change.

\\\ BATTERIES:
/// ENERGY STORAGE

The use of energy storage technologies, with our current paradigm being batteries, is essential to the success of the 4th industrial revolution.

- All the tech - IoT sensors, blockchain, AI, etc. - must have a source of power.
- As we begin to rely less on energy manufacturing and traditional utility grids and focus more on energy harvesting, we'll need to have better storage of just-in-time energy harvesting for use at a later time.
- Mobile technology requires mobile energy.

As a formal definition, a ***battery*** is a source of electrical power for powering electrical devices such as flashlights, computers, and automobiles. In the case of energy storage, batteries give us the ability to balance demand (energy when we need it) and supply (energy when we can capture it.)

Consider solar power. Today it's bright and sunny and your PV systems capture a lot of energy, so much energy that you can't use it in a normal day of operating your house. You need a way of storing that energy because tomorrow may be cloudy and rainy.

For that reason, we need to think about the intermittent supply associated with energy harvesting. Wind doesn't happen daily. For generating energy, sometimes we get too much, sometimes we get too little. The same is true for sunlight. We have to use the sunny days to save for the rainy days.

And of course, given that so much tech is going to be everywhere in the world, we cannot rely on the fact that the tech will be close enough to "plug into a wall outlet."

Types of Batteries

Energy inside a battery is stored as chemical energy (reactants, electrons, positive/negative electrodes, and all that ~~cool~~ stuff you learned in your high school chemistry class).

According to the chemical structure used inside a battery to store energy, you get one of two main types of batteries:

- ***Primary (non-rechargeable) battery*** - a battery whose chemical structure is not designed to be returned to its original chemical state. Thus, the battery is considered a one-time-use battery. Batteries in (non-smart) watches, key fobs, hearing aids, etc. are primary batteries. (A lot of these are commonly referred to as button batteries.) An alkaline battery, the most popular type of battery in the world, is technically a primary battery. (See the discussion on alkaline batteries below for rechargeables.) Alkaline batteries are used in children's toys, flashlights, TV remote controls, etc.

- ***Secondary (rechargeable) batteries*** - a battery whose chemical structure is designed to be returned to its original state by - you guessed it - recharging it. A lithium-ion battery is the most common secondary battery you hear about. Lithium-ion batteries are used extensively in tablet computers, laptops, electric vehicles, power tools, etc.

Alkaline Batteries

Alkaline batteries account for about 80% of all batteries worldwide. Alkaline batteries store energy in chemical form as a reaction between zinc metal and manganese dioxide. (We bet you're glad you know that.) They come in a variety of types and sizes including A, AA, AAA, C, D, and many other alphabetic configurations.

Can you recharge an alkaline battery? The short answer is yes, the longer answer is, well, longer. Some alkaline batteries are designed, manufactured, and advertised as rechargeable alkaline batteries. They are often green or light blue in color such as those offered by ULINE, Pale Blue, and Brightown. Of course, you have to buy a special recharging unit for these.

The trend is toward rechargeable alkaline batteries. IKEA, among others, no longer sells non-rechargeable batteries in their retail stores.

Experts agree that you shouldn't recharge an alkaline battery that isn't advertised as rechargeable for a variety of reasons.

You can and should recycle alkaline batteries. Many U.S. states are beginning to require the recycling of alkaline batteries, as opposed to throwing them in a landfill.

Lithium-Ion Batteries

A ***lithium-ion battery*** is a rechargeable (secondary) battery in which lithium ions move from a negative electrode to a positive electrode when you extract the energy. When you recharge a lithium-ion battery, the lithium ions move in reverse from the positive electrode to the negative electrode, and thus the battery is returned to its original fully charged state.

Lithium-ion batteries are all the rage in portable electronic devices (smartphones, laptops, tablets, etc.) and especially electric and hybrid electric vehicles (called, obviously enough, an ***electric-vehicle battery***, or ***EVB***). They are also popular in battery-powered tools such as drills, skill saws, weed whackers, lawn mowers, vacuum cleaners, etc.

You also see lithium-ion batteries used as a primary storage system for wind energy and PV solar systems. When your PV solar system captures the sun's energy and converts it into electricity, that electricity is most often stored in a lithium-ion battery (more appropriately, a lithium battery bank).

Solid State Batteries

It's pretty simple. We need batteries that are smaller, hold more energy, have shorter recharging times, have longer life spans, are environmentally friendly when thrown away (or recycled), and are not susceptible to leaks, explosions, and the like.

One such promising development is called a solid-state battery. A ***solid-state battery*** utilizes solid electrodes and electrolytes instead of liquid and polymer gel electrodes and electrolytes like those used in current lithium-ion batteries.

While right now expensive to make and still in their development, solid-state batteries hold the promises of:

- Higher energy densities (storing more energy in a much smaller space)
- Lower risk of catching fire because material is solid-state and not liquid polymer or gel
- Faster recharging
- Higher voltage
- Longer life

No doubt, batteries of the future will look and be much different from those of today.

Hot Companies to Watch in Battery Storage

Of course, Tesla will come to mind for most of you. Tesla, to be sure, is a leader in industry disruption not only in the automobile industry but also renewable energy. Tesla's Powerwall, for example, captures solar energy using photovoltaic (PV) systems and stores it in a "wall" or bank of lithium-ion batteries. You can use the wall of energy to power your house, recharge your electric vehicle, and as a backup power source.

Tesla, LG Chem, and Enphase Energy are the big three in the battery storage market. Together, they account for 85% of all sales in the battery energy storage market. [13]

But this is becoming a highly competitive space, with lots of new startups emerging all the time. Some of those include SK Innovation, Quantum-Space, Ionic Materials, and NEI Corp. [14]

\\\ ENVIRONMENTAL, SOCIAL, /// AND CORPORATE GOVERNANCE

Environmental, social, and corporate governance (ESG) is the new standard framework for business operations around the world and represents the 3 main topics that businesses are expected to report on. The purpose of ESG is to have companies capture and make available to the public all non-financial risks and opportunities inherent to the company's day-to-day business activities.

ESG is becoming not only a common practice for businesses, but in many cases is required. Just about every company you can think of publishes their ESG framework and practices every year in a variety of reports. Our focus for Energy Technologies is obviously E (environmental). Simply look up a company followed by ESG to see how it is reducing carbon emissions, going green, and more.

When considering a company for employment, consider its ESG practices. Is it doing enough? Not enough? Could it be doing things differently? Think about these things as you research different companies. There are even entire industries dedicated to ESG practices. It's up to you to continue using these 4th IR technologies to help companies become more energy friendly.

CHAPTER 9

Sensing Technologies

Hearing, Seeing, Feeling, and Smelling

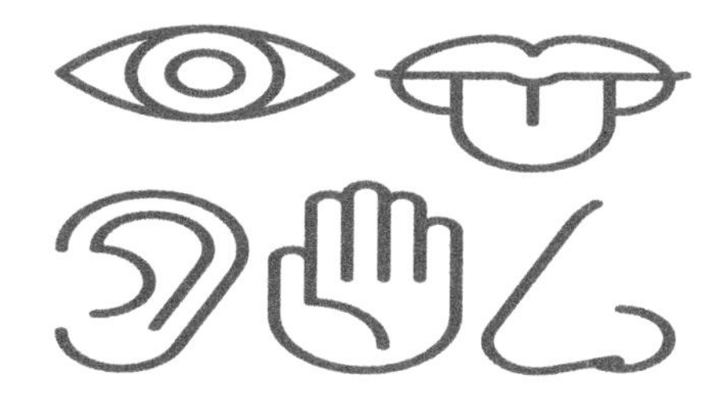

How many senses do we have?
A. 3
B. 4
C. 5
D. 6
E. 7

Another trick question, of sorts. Most people think we have 5 senses, but we actually have 7.

The 5 common ones are - of course - sight, sound, touch, smell, and taste. The two lesser-known senses are:

- *Vestibular* - (movement) the movement and balance sense, which gives us information about where our head and body are in space. Vestibular sensing helps us stay upright when we sit, stand, and walk. (Gymnasts have a keen vestibular sense.)
- Proprioception - (body position) the body awareness sense, which tells us where our body parts are relative to each other. Proprioception sensing helps us understand how much force to use, allowing us to do something like crack an egg while not crushing the egg with our hands.

Many hearing, seeing, feeling, and smelling technologies have been around for several years in the 3rd industrial revolution. We're not presenting them here specifically as technologies of the 4th industrial revolution. But we will need them in 4th industrial revolution innovations.

Sensors, Actuators, and Transducers
It's worth a bit of our time to talk about some technical terms before we jump into the sensing technologies of hearing, seeing, feeling, and smelling. Those terms are sensors, actuators, transducers, and a control module.

Let's consider an example of a microphone and a speaker. You speak into the microphone and your voice is amplified through a speaker. It's simple enough to think about doing karaoke with your friends. You sing into a microphone and your voice comes from the speaker. It's the type of thing we take for granted every day while using technology.

When you speak into a microphone, your voice is captured by a ***sensor***, a device that monitors and measures the physical aspects of an environment. In the case of karaoke, the microphone is listening to and capturing the sound waves of your voice.

Those sound waves are a form of energy, and they must be converted into an electrical-signal equivalent by a transducer. A ***transducer*** converts one form of energy into another form of energy. The control module of the karaoke machine cannot work with sound waves; it can only work with the electrical-signal equivalent of the sound waves. (Think of a control module as the brains or CPU.) So, the transducer takes the sound-wave forms of your voice (captured by the microphone), converts them into their electrical equivalent, and sends them to the control module.

The **control module** then passes the electrical signals of your voice to the speaker. For the speaker to reproduce your voice, it must first convert the electrical signals of your voice (one form of energy) into sound waves (another form of energy). This, again, is achieved by a transduction process, the process of converting one form of energy to another form of energy via a transducer and actuator. An ***actuator*** takes the electrical signals from the control module, uses a transducer to convert the electrical signals, and then produces another form of energy. In the case of karaoke, the actuator is the speaker which amplifies your voice.

In short, a sensor monitors what is happening. A transducer takes what "happening" has occurred and converts it into electrical signals. The control module more or less manages everything and processes the electrical signals, which may include passing the electrical signals to an actuator. Within the actuator, a transducer converts the electrical signals into what the actuator needs for the appropriate output. The actuator then creates the output. The output can be a sound, as is the case with karaoke, or it could be a light flashing, the causing of a motor to rotate, a text message being sent, etc.

(Hope you followed all of that.)

Suffice it to say that many people don't use the technical terms of sensor, actuator, transducer, and control module correctly. And that's okay. The technical details aren't often that important. What is important is understanding the capability of the technologies. Where possible, we'll attempt to use the right technical terms. However, please understand that sensor is most often used (incorrectly) as the term for also a transducer and an actuator.

\\\ HEARING /// TECHNOLOGIES

Hearing technologies are those technologies that detect (and perhaps) interpret sound, which could be the sound of a door closing, your voice, a simple thump, or even the sound termites make while eating the frame of your house.

IoT Sound Sensors

These are microphone-based sensors that detect the presence of sound or noise and usually cause some sort of response (handled by an actuator). Sound sensors are popular for applications like home security systems and monitoring, for example, turning on the outside lights when sound is detected. (This can also occur via a movement sensor.) They are also popular for basic control of home appliances like lamps (clap once for one, twice for off).

They differ from advanced speech recognition, which not only detects sound or noise but also must capture sound at a certain quality to make sense of words.

For IoT sound sensors, you can choose from among a wide variety of microphone types like dynamic, carbon, ribbon, and condenser. You'll need to select the microphone type based on the situation and what you're trying to capture. (We'll let you research more on different types of microphones at your leisure.) You can also choose from characteristics such as big or small (size). And you can control the sensitivity of an IoT sound sensor. IoT sound sensors that you program to be less sensitive will only detect louder sounds.

Basic IoT sound sensors are about $1 in price.

Speech Recognition

Speech recognition, often called ***automatic speech recognition*** or ***ASR***, are technologies that detect, interpret, and respond to your spoken words. Obviously, things like Amazon Alexa, Google Assistant, and Siri rely on speech recognition technologies, as does your remote control so you can say things like "Sopranos" instead of using that long arduous process of moving the cursor around on a keypad on your TV, computer, or tablet screen.

Many such hearing technologies incorporate machine learning. The more you speak to a hearing technology, the more it learns to recognize not only your voice but also your tone, inflection, and use of certain words.

Speech recognition is also vitally important as an assistive technology for people with sight or movement limitations.

If you want speech recognition for your personal computing needs, you'll find several great ones in the cloud. (You might want to do a little reading on cloud computing in the Infrastructure Technologies chapter.) These include: [1]

- Google Cloud Speech API
- Microsoft Azure Cognitive Services
- Amazon Transcribe
- IBM Watson
- SpeechMatics

As with all cloud or as-a-service offerings, they usually charge by the minute or word.

Variations of speech recognition are present in many aspects of your life. Some elevators respond to verbal commands… open door, 3rd floor, etc. Music apps like Shazam, SoundHound, and Musixmatch all use a variation of speech recognition. Language translation platforms typically make extensive use of ASR technologies.

\\\ SEEING /// TECHNOLOGIES

The formal term for seeing technologies is computer vision. In short form, ***computer vision*** is all about image recognition and understanding by a computer. Computer vision is our attempt to replicate the ability of the human visual system. So, computer vision includes such things as the basic capturing of image and video all the way up through using artificial intelligence to determine what an object is (see convolutional neural network), how far away it is, how fast it's moving, etc.

(Seeing technologies also include the "production" of things for you to see. These include the simple and straightforward like your computer screen and also the more advanced like headsets and visual displays for augmented, mixed, and virtual reality environments. And, as we discussed at the beginning of this chapter, these seeing "production" systems might very well include transducers and actuators.)

Lots of important technologies in the "seeing" realm. Just a few are listed here, all of which are vitally important to the success of the 7 core 4th industrial revolution technologies.

360 Camera

A ***360 camera***, also called an ***omnidirectional camera***, has a 360-degree seeing ability, thus giving it the ability to capture both still photos and videos of the entire surrounding area. 360 cameras are popular for applications such as:

- Surveillance and security - in-home security cameras, for example, that can detect motion.
- Google Street View - uses 360 cameras to give you the ability to see in different directions while in Street View.
- Virtual Reality - 360 cameras are popular in this venue because of your need to move your head and see in different directions. 360 cameras used for this purpose are often called *VR cameras*, special types of cameras that render a 3-dimensional view specifically for VR applications (and perhaps mixed reality as well).
- Live event (virtual attendance) - not much need to explain this.
- Real Estate - for enabling potential buyers to virtually tour a home, building, etc.

Sonar, Radar, and Lidar

These all work on the same basic principle; they measure the time it takes for a signal to reach an object, bounce off that object, and return to the sending unit. They differ in that sonar uses sound, radar uses radio waves, and Lidar uses light waves.

- ***Sonar*** - **So**und **N**avigation **a**nd **R**anging
- ***Radar*** - **Ra**dio **D**etection **a**nd **R**anging
- ***Lidar*** - **Li**ght **D**etection **a**nd **R**anging

Each has their advantages and disadvantages, making them more suitable in certain situations and not in others.

For a long time now, sonar has been the dominant "seeing" technology in the water. Sonar is used by navies to detect vessels (both submarines underwater and top-water vessels). Sonar is also used to map the ocean floor and search for underwater objects. When you read about sonar technologies, you'll see a couple of other terms pop up… ultrasound and ultrasonic. These are subsets of sonar technology. For example, ultrasound is used in the medical field to "peek" inside the human body.

Radar is popular in other realms including spotting aircraft in the sky (air traffic controllers) and determining the speed of moving vehicles (law enforcement).

Lidar is all the talk for autonomous vehicles. Lidar emits millions of laser light pulses to create a 3D map of the surrounding area. It has an extremely high degree of accuracy. See Chapter 6 on Autonomous Vehicles for more on the use of sonar, radar, and Lidar.

With the release of the iPhone 13, Apple began putting Lidar capabilities in its phone. (The other competitors in the phone space have as well.) With Lidar on your phone, you can, for example, map the exact layout of a room. You can capture height, length, width, depth, distance, and so on.

It's going to be fascinating to see the innovations within phone-based Lidar. For example, you could map a room and then add falling snow. The snow would accumulate in various ways on the furniture and floor but not the floor underneath the coffee table. Why? Because the Lidar mapping would note surface area above the floor which would prohibit snow from falling there.

OCR

Optical character recognition (OCR) is the electronic conversion of typed, handwritten, or printed text into a computer-usable format. Optical character recognition has been around for quite some time, dating back into the 3rd industrial revolution, but remains an important technology for the 4th industrial revolution.

Applications and uses of OCR include:

- Assistive technologies for the sight-challenged (e.g., converting printed text into audio format)
- At-register scanning and processing of checks
- Photo-based depositing of checks
- Traffic sign recognition (important for autonomous vehicles)
- License plate recognition (automated ticketing systems for driving violations and also toll roads)
- Scanning printed documents to create searchable PDFs
- Handwriting recognition (accompanied by machine learning techniques)

SLAM

SLAM (simultaneous localization and mapping) is a seeing technology for environments that need the constructing and updating of an environment including the location of the "agent" within it. Here, think about the popular iRobot Roomba, the self-driving robot vacuum cleaner. It uses SLAM to map a room, detect objects to avoid, and keep track of where it is in the room. The "agent" is the iRobot Roomba.

Biometrics

The use of biometrics is gaining more societal acceptance daily. ***Biometrics*** literally means the measures of life, a combination of "bio" or *life* and "metrics" or *measures of.* Biometrics can be something as straightforward as taking your blood pressure or advanced applications like using your fingerprint for identification and really advanced applications in biomedical engineering, bio-technology, and neuro-technology. Biometrics also touches on 4th industrial revolution applications like using 3D printing to create synthetic skin grafts (and someday 3D-printed fully artificial organs).

As we begin to see more of the integration of technology and the human body, biometrics will become increasingly important.

Motion Capture

Motion capture (also ***mocap*** or ***mo-cap***) is the process of detecting and recording the movement of objects or people.

Motion capture is used for a variety of applications. In filmmaking and video game development, motion capture is used to capture the movement of actors to create 2D and 3D animated renderings. In the field of physical therapy, clinics use motion capture to determine a patient's progress. In sports safety, researchers use motion capture to understand, for example, how the body changes from impact to design better safety and sports equipment such as helmets, safety harnesses, and the like.

Subsets of motion capture include gesture recognition, hand and finger tracking, and eye tracking.

Gesture Recognition

Gesture recognition is a technology with the goal of recognizing human gestures. Human gestures can include the shrugging of shoulders, changes in facial patterns due to emotional changes, in-the-air swiping by the hand for touchless interfaces, body movements while using virtual weapons while in a virtual reality game, and so on.

An important part of gesture recognition is hand and finger tracking.

Hand and Finger Tracking

Hand and finger tracking is a type of motion capture for detecting and understanding finger and hand movements.

One of the first applications of hand and finger tracking was in the interpretation of sign language. Evalk, for example, offers GnoSys, a smartphone app that can translate sign language into speech. [2]

Hand and finger tracking take place in 2 environments. The first is like GnoSys, which doesn't require the use of gloves for capturing hand and finger movement. It, instead, uses video capture. The second is the use of gloves to track hand and finger movements, which fall into the category of feeling or haptic technologies.

Eye Tracking

Yet another part of gesture technology is ***eye tracking***, the process of measuring the point of gaze, i.e., where someone is looking. Eye tracking technologies are particularly helpful in human-to-computer interaction (HCI) for people with movement limitations that prohibit them from using traditional input devices like keyboards, mice, and trackpads.

Eye-tracking technologies are becoming increasingly prominent in marketing and customer experience management. Eye-tracking technologies enable marketers to detect what captures a consumer's attention, in what order a consumer reads information on a box, and so on.

You may have even taken a test using a browser with eye-tracking technology to prevent cheating.

\\\ FEELING /// TECHNOLOGIES

The formal term for feeling technologies is haptic technology. ***Haptic technologies*** are those technologies that (1) create the sense of touch by applying force and motion (including vibration) to a person, or (2) detect and interpret force or motion from a person. The computer is sending you "feeling" in the form of haptic interfaces and/or you are sending the computer "feeling" in the form of your motions, actions, etc.

When I "Feel" the Computer
There is a big range of devices in this area. You can have a game controller that vibrates to give you a sense of some sort of explosion. You can also use a specialized gaming chair that provides vibration.

Many automobiles are incorporating these types of technologies as well. For example, when you start to change lanes and your car notes that another car is in the other lane, your car may vibrate the steering wheel or even make it difficult for you to turn the steering wheel in that direction.

Even putting your phone on vibrate is about the use of a haptic technology.

(These haptic technologies - when the computer is sending you "feeling" - are achieved through a combination of a transducer and actuator.)

When the Computer "Feels" Me
The easy (and dull) ones include things like using a mouse or touchpad to move the cursor on the screen, clicking on buttons, highlighting, and the like and also moving a joystick and selecting buttons on a controller. It can also include:

- Hand controllers in extended reality environments for basic interface functions and also specialized functions like using a bow and arrow.
- Headsets - electronic headsets that adjust their field of vision to your head moving in different directions, up and down and side to side. Headsets are actually a combination of feeling (capturing and interpreting your head movements) and seeing (changing your field of vision) technologies in virtual reality and mixed reality applications. (See Chapter 4 on Extended Reality for more on virtual and mixed reality applications.)

When the Computer "Feels" Other Things
Feeling technologies are not limited to computer-to-person interaction. You can apply feeling technologies in a wide variety of environments. These would include things like:

- Vibration/Motion sensors - placed on equipment to determine when a machine may be coming out of balance.
- Pressure sensors - to determine weight change, for example.
- Tilt sensors - to determine the angular orientation of something (e.g., is a platform for a wind turbine starting to shift?) Tilt sensors support the vestibular sense.

An interesting application in this space is that of a digital twin. A ***digital twin*** is a real-time digital counterpart of a physical object. Take the wind turbine as an example. To a wind turbine you can add IoT sensors all over it (and around it) to create a real-time digital version of it. The IoT sensors can measure stress, motion/vibration, speed of the wind, stability of the platform, changes in the soil structure around the turbine, and so on. This enables you to monitor the wind turbine without being at the wind turbine location to constantly monitor it physically.

\\\ SMELLING /// TECHNOLOGIES

Smelling technologies (also known as ***digital scent technologies*** or ***olfactory technologies***) are technologies that can detect natural scents/smells/odors/molecules/particles in the air and also produce synthetic scents/smells/odors/molecules/particles. Obviously, smelling technologies attempt to replicate the human olfactory system.

Scents, smells, odors, molecules, and particles… may seem like a weird list, so let's explore a bit.

Your sense of smell detects the presence of odorous molecules in the air. The binding of odorous molecules with your olfactory receptor neurons creates a chemical stimulus causing electrical signals to be sent to your brain.

It is this sense of smell that allows you to smell when food is going bad, perfume, dirty socks, flowers, and someone's bad breath. These are about odors, scents, and smells, which result from a change in the molecular structure of the air around you. These are the types of odors (i.e., molecular structural changes) that you can "smell."

Other types of molecular structures cannot be detected as "smells." The most common of these in our homes is carbon monoxide. Carbon monoxide is a colorless, odorless, and tasteless gas. That's why you have carbon monoxide detectors in your home; you can't see or smell carbon monoxide, so your carbon monoxide detector recognizes their molecular presence in the home. Natural gas is another gas that has no odor. Companies that sell natural gas add a harmless gas called mercaptan to natural gas to give it that rotten egg smell.

Finally, other types of smelling technologies use the presence of particles in the air to detect "smell." The most common of these is a smoke detector in your home. Smoke produces particles in the air, which smoke detectors can detect or "smell."

In this smelling space, we have lots of IoT sensors. There are IoT sensors specifically for smelling carbon monoxide, carbon dioxide, and natural gas. You can program a general smelling sensor for things like food going bad or the presence of some diseases, infections, abscesses, or tumors that have an external manifestation (i.e., not internal). And, of course, there are smoke detector sensors which detect smoke-related particles in the air.

That's the detection side. There are also smelling technologies that produce synthetic smells. You should check out OVR Technology, which has created an Architecture of Scent Platform. Developers can use this platform to create a more immersive experience through the sense of smell. Visit https://ovrtechnology.com and https://www.youtube.com/watch?v=dYilOb5lIog. Absolutely fascinating.

By the way, your olfactory sense has a stronger link to your memory than any of your other senses. So, if we want to create technology-based memorable experiences, we'll need to include the sense of smell via technology.

Barcodes (UPC and QR codes)

Barcode (also ***bar code***) is a method of representing data in a machine-readable, standardized format. You've seen these all over the place, mostly as either UPC or QR code.

A ***universal product code (UPC)*** is a type of barcode that uses vertical bars and the distance between them and their thickness to determine a number. These are referred to as linear or 1D (1-dimensional) barcodes. A UPC has 12 numbers, with the most common application being a product number in an inventory system. Think of just about any product in a grocery store. It contains a UPC that uniquely identifies the product, such as a box of Cheerios. Of course, the limitation is that every box of Cheerios has the same UPC, so the store can't distinguish between each box.

A ***QR (quick response) code*** is a matrix barcode and referred to as a 2D (2-dimensional) barcode. QR codes can contain much more information than a UPC. Many QR codes commonly contain date and time information.

Organizations can and do create their own unique QR codes, often for one-time use. You've undoubtedly received a QR code for event tickets, promotions, etc. These were generated by the organization (e.g., SeatGeek, TicketMaster, or StubHub) and literally discarded after the specified use. Of course, QR codes can be more permanent. During the COVID pandemic, restaurants stopped distributing menus. Instead, patrons scanned a QR code at their table to view a copy of the menu online.

There are many different and interesting configurations of QR codes including Aztec Code, SnapTag, and SPARQCode. Explore them on your own and "nerd out" if you want.

CHAPTER 10

Communications Technologies

Anywhere Is the New *Where*

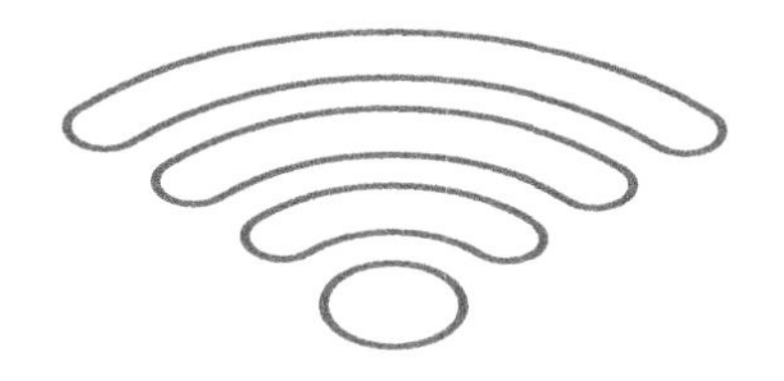

Faster, better, and more reliable communications technologies will enable more widespread adoption and use of the key core 4th industrial revolution technologies including autonomous vehicles, drones, IT, and extended reality, as well as all the others.

Like sensing technologies, many communications technologies have been around for several years in the 3rd industrial revolution. We're not presenting them here specifically as technologies of the 4th industrial revolution. But we will need them for the 4th industrial revolution innovations.

Bluetooth

Bluetooth has been around for quite some time, dating back to its first commercial uses in the early 2000s. ***Bluetooth*** is a short-range wireless technology that has been standardized for communications between devices (both fixed and mobile) over short distances, typically up to about 30 feet. For you the most common applications include connecting your headphones/earbuds to your phone, connecting your phone to your computer, and connecting your phone to your car's entertainment and navigation systems.

Bluetooth is essential to the success of the 4th industrial revolution, especially in the IoT space. Bluetooth boards are relatively small, some as small as about 1-inch square. And they are relatively inexpensive, with some costing as little as $6. Bluetooth is great for connecting IoT sensors to each other and to a control unit that manages the IoT sensors and collects and processes information from the sensors. Consider, for example, sensors buried in a lawn that communicate with a water sprinkler control system so it can adjust watering frequency and lengths of time based on the dryness of the soil in various parts of the yard.

Bluetooth has many variations. The two main ones are classic Bluetooth and Bluetooth LE, with LE standing for *low energy*. Classic Bluetooth is widely used in situations for which you need a constant connection and move a lot of information. Wireless speakers would be an example. Bluetooth LE is for situations when you only need an intermittent connection and not move a lot of information. Tile and all its competitors, the popular solutions for *where's-my-fill-in-the-blank*, use Bluetooth LE. Many IoT applications use Bluetooth LE.

To be sure, Bluetooth LE is smaller and costs less than Classic Bluetooth. If your innovation requires a Bluetooth type connection, you'll want to learn more about the different types of Bluetooth.

Radio-Frequency Identification (RFID)
The wireless connectivity in your credit or debit card that allows you to make contactless payments most likely uses NFC, near-field communication, which is a subset of RFID. So, let's talk about RFID first and then NFC.

Radio-frequency identification (RFID) is a communications technology that uses electromagnetic radio waves (sent by an RFID reader device) to identify and track tags attached to objects. The process is quite simple and widely used. The reader device sends out a radio wave signal and the RFID chip responds by sending back digital data, which at a minimum is some sort of identification, to the device reader. This identification can be a unique tag number (which would be used to access a database of information about the object), an ISBN (for a book), badge number, an inventory UPC, your credit card number, etc. Some RFID tags transmit other information such as product lot number and date of production (for product inventory), expiration date and security code (for a credit/debit card), and so on.

RFID tags are (1) either passive or active, (2) either beacon or transponder, and (3) either read-only or read-write (and, obviously, various combinations of the three types).

Passive RFID tags have no battery, but instead use the energy from the electromagnetic radio waves to create enough electrical energy to cause the RFID chip to send out its digital data. Passive RFID tags work well for short distances (usually up to a few feet) and are inexpensive in bulk, often as low as $.05.

Active RFID tags have a built-in power supply and work over greater distances and can hold and communicate more information than passive RFID tags. Active RFID tags can work over distances of several hundred meters. They also cost upwards of $25 or more.

Beacon RFID tags continuously send out information at a predetermined time interval. These are often used in real-time locating systems. Transponder RFID tags, on the other hand, only send information when they receive a radio signal from the reader device.

Finally, RFID tags can also be either read-only or read-write. A read-only RFID tag can only transmit its digital data. You cannot send information to a read-only RFID tag and have it update the stored information. Of course, read-write RFID tags have both the ability to send digital data and to receive digital data and update the information stored on the tag.

RFID tags are widely used.

- Libraries - to check-in and check-out books
- Livestock tracking
- Passports
- Typical product inventory management (retail shelves, warehouse applications, etc.)
- Human implants (yes, this is happening)
- Access control (badge identification)
- Healthcare
- Transportation (toll roads, for example)

Near-Field Communication (NFC)

Near-field communication (NFC) is a subset of RFID but has data and transmission speed and distance limitations. Most all NFC applications are limited to a few centimeters, usually no more than 10 (approximately 4 inches).

Your credit or debit card probably uses NFC. The technical term for that particular payment communications standard is ***EMV***, which stands for **E**uropay, **M**astercard, and **V**isa. Those companies worked together to develop EMV and were the first to roll it out in the payments space.

The concept is the same for NFC as for RFID. When you "tap" the reader device with your credit card, your credit card is close enough to the reader device that it activates your EMV which in turn transmits your credit card information to the reader device. (BTW, the physical "tapping" is of no importance. Contact not required.) So, EMV-supported NFC tags are passive (they have no power supply), of a transponder type (they only send information when they receive a radio signal), and read-only (they only send information and do not update information within the tag).

WiFi

WiFi (wireless fidelity) is a suite of wireless network protocols that support local area networking for devices and Internet access. (We most commonly use the term WiFi, but its technical term is Wi-Fi. Perhaps we got tired of typing that silly dash all the time.)

Not much to say here, kind of like Bluetooth. WiFi is the most commonly used network infrastructure in the world, especially for homes, offices, coffee shops, airports, etc. Network range is typically about 75 feet but can range up to 500 feet, especially in outdoor applications.

LiFi
(Finally, something new and interesting)
WiFi will continue to get faster, better, and more reliable. And we think we'll be using WiFi for the foreseeable future, but LiFi is on the horizon. ***LiFi*** (***light fidelity*** or ***Li-Fi***) is a wireless communications technology that utilizes light to transmit data, as opposed to WiFi that uses radio frequency waves. LiFi can transmit data at extremely high speeds using visible light, ultraviolet, and infrared.

The promise of LiFi is substantial. Firstly, it can transmit data in excess of 100 gigabits per second. (As of early 2024, the average speed for a home WiFi unit was less than 1 gigabit per second, and more typically around 100 to 200 megabits per second.) Secondly, because LiFi uses light transmission and not radio frequency waves, it would not interfere with other RF-based networks. Think about being on an airplane. A LiFi network wouldn't interfere with the communications network of the cockpit, and you could literally turn on your overhead light to boost your speed.

Someday, you may pick out a ceiling fan for your home and choose your lights based on how much you want to boost your network speed. So cool.

Currently in the invention state but watch the development closely. For about $2,000, you can get a LiFi kit from LiFi.co. Check it out at https://lifi.co/lifi-product/lifimax-flex/.

The Gs, 2 through 5
The G in 2G, 3G, 4G, and 5G refers to generations in cellular or mobile communications. (Hopefully, you're well beyond 2G and 3G, but they still do exist in some areas.) Let's cover a couple of terms quickly and then we'll compare 2G, 3G, 4G, and 5G.

- ***Cellular network*** - communications network that wirelessly connects land areas called "cells," with each cell having at least one transceiver. A transceiver is a device that can transmit and receive data to and from wireless devices and to and from other transceivers. When you're out and about using your phone, you're doing so using a cellular network.
- ***Broadband*** - a type of data transmission that supports multiple types of communications sent simultaneously at very high speeds. Broadband is supported by wireless connectivity (cellular network, satellite, microwave, etc.) and also land-based connectivity such as optical fiber.
- ***Broadband cellular network*** - when you combine the 2, you get a broadband cellular network with the main focuses being (1) connected *cells* or land areas, (2) technologies enabling wireless connectivity, (3) the

movement of multiple (perhaps hundreds of thousands) communications simultaneously, and (4) very high speeds.

In general, each of the major mobile phone service providers has its own cellular network. These would include service providers like AT&T, Verizon, and T-Mobile. Smaller mobile phone service providers such as Mint, Cricket, and Boost don't own their own cellular network but rather lease capacity from the major mobile phone service providers. Each of these smaller service providers is called a *mobile virtual network operator (MVNO).*

Okay, back to the Gs. In short, each new generation in mobile communications brings faster speeds [1]

- ***2G*** - 64 kbps (the 1990s, probably before you were born)
- ***3G*** - 8 mbps (early 2000s)
- ***4G*** - 50 mbps (came out in 2009)
- ***5G*** - 10 gbps (introduced in 2018)

Hopefully, you have 5G capability. As of early 2024, 5G was still all the advertising rage. It covers roughly 62% of Americans, but that number is a bit deceiving. "Coverage" in this case means being available in the homes of 3 out of every 5 Americans. Land coverage is much, much lower. So, you may have 5G while you're in a populated land area, but quickly lose it once you travel into more rural areas.

The hope (hype?) with 5G is that we can use our wireless devices to enjoy things like high-quality on-demand video streaming and virtual reality with minimal or no latency. ***Latency*** is the delay that occurs between your action, for example moving your head in a VR game, and the game's response by changing your field of view.

6G

6G is the sixth-generation standard for broadband cellular networks, which every cellular phone provider plans to use. 6G will have significantly faster download speed and really promises to enable the use of technologies like virtual/mixed reality, IoT, and AI over a broadband cellular network without any latency or noticeable degradation in speed.

6G is coming, probably around 2030. Expect it to be upwards of 100 times faster than 5G, or about 1,000 gbps.

7G… ha! But stay tuned.

Microwave and Satellites

These are communications infrastructure technologies that enable the transmission of large amounts of information and communications simultaneously at extremely high speeds, usually over fairly significant distances.

Microwaves and ***satellites*** are line-of-sight communications technologies that transmit massive amounts of data over great distances in a very short period of time. These really are the enabling infrastructure technologies that Internet service providers, TV stations, etc. use to move massive amounts of data around the world.

These are line-of-sight technologies because the sender and receiver units have to be able to see each other. That's different from your home WiFi network. Signals inside your home from your wireless router to and from your computer bounce off of walls. So, you can be just about anywhere in your house with your computer and still have access to your wireless network.

Microwaves and satellites use a different transmission methodology, such that the signals don't bounce around. You've seen microwave relay stations and towers. They are positioned about every 8 to 10 miles to enable communications around the curvature of the earth. Satellites are similar except that they are orbiting above the earth. The sending and receiving satellite dishes on the ground must be in line of sight to the satellite.

A great development to watch in this space is Starlink, started by Elon Musk. The goal is to provide Internet access via satellites to the entire global population. There are currently almost 5,300 LEOS (low-earth orbit satellites) in the Starlink system, with each connected to dedicated ground transceivers. The full deployment of Starlink may include as many as 42,000 LEOS.

Most Innovations Require Multiple Communications Technologies

Suppose, for example, that the "smelling" sensor in your refrigerator detects that the milk is going bad. It could use Bluetooth to communicate that information to your refrigerator's central command center (oh, fancy) which would then use WiFi to connect to your smart home system which would order milk (and send you an alert text message… bad milk). That order would travel from your WiFi through your 5G network provider to the grocery store. Within your 5G network, communications technologies such as microwaves may be used.

This is an example of the concept of ***tethering***. In this case, you're tethering together all these communications technologies to essentially enable your smelling sensor in your refrigerator to communicate with the grocery store.

Wired Technologies

The communications technologies we've discussed here are obviously all wireless. And wireless is key to mobility. Mobility, enabled by wireless communications technologies, is all about the new *where*, specifically anywhere.

Of course, wired technologies are still important. Technologies such as fiber optics will continue to play an integral role in the success of technology.

CHAPTER 11

Infrastructure Technologies

Quantum, Cloud, and Nano

\\\ QUANTUM /// COMPUTING

Google announced it in 2019; it had achieved *quantum supremacy*. Its quantum computer solved a mathematical computation in 3 minutes and 20 seconds that would have taken the world's current fastest computer over 10,000 years to solve. [1]

That's the hope and hype of quantum computing… unbelievable speed.

About the time of Google's announcement, the fastest supercomputer in the world, named Fugaku (a joint development of RIKEN and Fujitsu), was capable of performing 425 petaflops. A petaflop is one thousand million million floating point operations per second. *Peta* is the term applied to 1,000,000,000,000,000 or 10^{15}. We measure computer processing speed in terms of how many floating point operations per second (abbreviated as *flops*) the computer can execute. So, Fugaku could perform 415,000,000,000,000,000 operations per second, and - at that rate - it would have taken Fugaku over 10,000 years to solve the mathematical computation that Google's quantum computer did in 3 minutes and 20 seconds.

Like we said, unbelievable speed.

To understand how and why quantum computers are so fast, we need to have a discussion about our classical (current) computer technologies and how they work.

Bits - Classical Computers

All information in classical computing technologies is stored, processed, and communicated in binary form. Binary means two. The smallest unit in binary is called a ***bit***, which stands for binary digit. Because binary means two, a bit can take on only one of two values, either 0 or 1. We use this binary concept in many ways… a door/gate is either open or closed, a light is either on or off, a coin flip results in either a head or tail… you get the idea.

Inside a computer, there is no single character for the letter P, the number 7, or an asterisk. Instead, we represent each of these (and all the other letters, numbers, and special symbols) as a unique series of 0s and 1s. The letter P is 01010000. The letter p (small p) is 01110000. An asterisk is 00101010. Again… you get the idea.

(Quick side bar and perhaps a little humor: A single binary digit is a bit. If you put together a logical group of bits to create, for example, the letter P, you create a byte. Bits and bytes. Interestingly enough, the technical term for half of a byte (4 bits) is a nibble. No kidding. And, half of a nibble (2 bits)? The technical term is a crumb. Those crazy computer people.)

Time for another question: We (humans) work with numbers in Base 10. So, for each digit position in a number such as 357, there are 10 possible digit representations, 0 through 9. Why do we work in Base 10 and not some other Base scheme such as Base 8 or Base 13? (Sorry, no multiple choice here. Think about this for a moment and write your answer below.)

Answer: ______________________________

It's not a trick question, but the simplicity of the answer seems almost too obvious. We work in Base 10 because we have 10 fingers. That's it. If we were 12-fingered mammals, we would be working in Base 12.

Let's consider the number 357. You can break that number down into its Base 10 equivalent as follows:

3 5 7

$7 \times 10^0 = 7 \times 1 = 7$

$5 \times 10^1 = 5 \times 10 = 50$

$3 \times 10^2 = 3 \times 100 = 300$

357

Each digit position can take on 1 of 10 possible representations, 0 through 9. We multiply the digit representation in each position by continually increasing the power of the base, starting with 0 on the right. (We're sure you remember all that from your elementary school math class. Chuckle.)

Binary works the same way, except that each digit position can take on 1 of 2 possible representations, 0 and 1, and the Base is 2. The number 357 in binary would look like this:

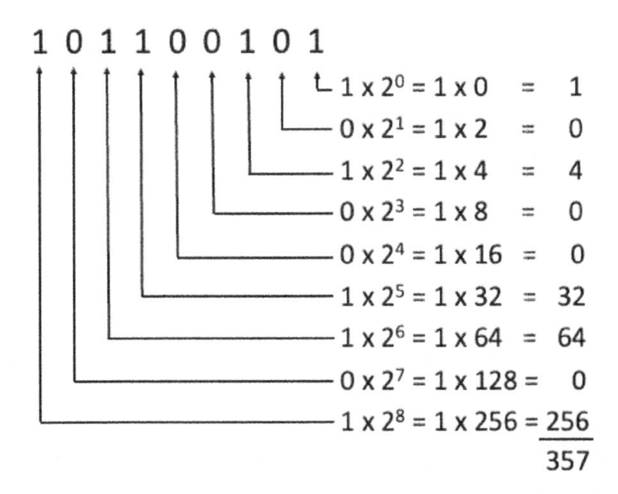

So, when you're working in Excel with the number 357, your computer is working with the equivalent of 101100101. It's quite fascinating when you think about it… the simplicity with which computers represent and work with information.

Qubits - Quantum Computers

But quantum computing is different from classical computing technologies. Quantum computing's most basic unit for representing information is a ***qubit***, which stands for quantum binary digit. What's really interesting and distinguishing about a qubit is that it can be both 0 and 1 at the same time. This simultaneous representation of 2 possible states is called *superposition* in quantum mechanics.

So, formally, let's say that a ***quantum computer*** (or ***quantum computing***) is an alternative to classical computing technology that offers unbelievable speed because it performs computations on information in quantum states such as superposition. There are other quantum states like interference and entanglement that also offer speed increases and other benefits over classical computing technologies, but now and here is neither the time nor the place.

The easiest way to think about a qubit and superposition is to turn a coin on its side and spin it. While the coin is spinning you can see both heads and tails, the possible states that result from a coin flip. For our discussion, we'll represent a qubit like this:

It looks kind of weird, but the notion is that a qubit can be both 0 and 1 at the same time.

The ability of a qubit to take on 2 values simultaneously is why quantum computing is so much faster than classical technologies. Let's illustrate with an example. Suppose we wanted to add all possible combinations of 2 classical bits and the same for 2 qubits. It would look like this.

CLASSICAL BITS				QUANTUM BITS			
0 +1 1	0 +0 0	1 +1 10	1 +0 1	1	0	+ 10	1
4 Computations				1 Computation			

Using classical bits and technology, you would have to perform addition 4 times. With qubits and quantum computing, you only perform the addition 1 time, with that singular arithmetic operation resulting in 4 answers.

Of course, you're smart and you're thinking… "Cool. Quantum computers can do things 4 times faster. Clearly though, 3 minutes and 20 seconds compared to 10,000 years is much more than 4 times faster. I don't get it." (Kudos for thinking that.)

In the above example of adding all possible combinations of 2 classical bits and qubits, the speed increase really isn't 4x, but rather 2^n where *n* is the number of qubits used. So, if you use 2 qubits, you get an increase of 2^2 in speed (4x). If you then want to add all possible combinations of 3 bits, you can get an increase of 2^3 in speed (8x). Rinse and repeat and you can start to see that speed doubles every time you add another qubit.

This is the notion of *scaling*. We're not adding to or even multiplying speed increases, we are raising the power each and every time we add another qubit.

Another illustration… scaling with a story about rice and the game of chess.

There was once a game maker who invented the game of chess. He took it to the queen, knowing that she loved to play games. After explaining the game of chess and playing it with the queen for a while, she wanted it and offered to buy it from the game maker.

His price was simple. For the lower left corner on the board, pay me 1 grain of rice. For the square next to it on the first row, pay me 2 grains of rice. For the square next to it on the first row, pay me 4 grains of rice. Rinse and repeat for each square on all the rows, continually doubling the amount of rice payment for each square.

The queen did a quick computation in her head for the first row:

- Square 1, 1 grain of rice
- Square 2, 2 grains of rice
- Square 3, 4 grains of rice
- Square 4, 8 grains of rice
- Square 5, 16 grains of rice
- Square 6, 32 grains of rice
- Square 7, 64 grains of rice
- Square 8, 128 grains of rice

That was for the first row. To pay for the first row of squares, the queen had to give the game maker only 255 grains of rice, much less than a handful. It sounded like a great deal to her, so she readily accepted the terms of payment.

Bad decision. Why? Because of scaling.

Think about the second row of 8 squares. The rice payment for each square is 256, 512, 1024, 2048, 4096, 8192, 16,384, and 32,768. We're already at 32,768 grains of rice for a single square, and we've got 6 more rows of 8 squares each to go.

No need to do the remaining rows and squares. But we can tell you this, in the end the queen had to pay for the chess game an amount of rice equivalent in size to Mount Everest.

That's the notion of scaling.

So, with each new qubit, we double the speed of quantum computing.

The Medical Realm
Medicine, health care, medical… whatever you want to call it… is perhaps the field of study and application that will benefit most from advances in 4th industrial revolution technologies. And quantum computing will dramatically impact and disrupt (in a good way) the field of medicine.

For example, the human genome - all genetic information in a person - contains 3 billion base pairs. Storing this information using classical bits requires approximately 1.5 gigabytes of storage. That same amount of information can be stored in 34 qubits. That's right, <u>34</u> qubits. And, if we double the number of qubits to 68, we would have enough room to store the complete genome of *every* person in the world. [2] Using classical technologies, the storage requirement equivalent is 12 million million million bytes, or 12,000,000,000,000,000,000.

Working with and researching the human genome are among the most complicated and time-intensive tasks not only in medicine but in everything we use computers for. Drug testing and trials are another area that will benefit from advances in quantum computing. Prior to receiving approval for human testing, pharmaceutical companies must run millions and millions of "simulations" to understand drug interactions, side effects, and so on. Sometimes, this can take years. But, with quantum computing…

The Artificial Intelligence Realm
As you learned in the chapter on artificial intelligence, many AI (1) use very large and complex sets of information to arrive at an answer, and/or (2) rely heavily on learning from vast amounts of information.

Using quantum computing, we can reduce the size of the information for arriving at an answer because we can store and process that information in qubits. The less information you have to consider for making a decision, the faster you can make the decision (generally).

And, we can greatly reduce the time it takes to train an AI, i.e., its learning. Instead of using classical bits to feed the AI one scenario at a time, we can simultaneously feed the AI thousands, if not millions, of scenarios at once using qubits. This is especially true in generative artificial intelligence.

The Security Realm - A Rather Serious Implication

While quantum computing promises great computational speed, there exist some significant downsides to being able to perform a task "so fast," especially in the security realm.

In the cryptocurrency and blockchain chapter, we talked briefly about the 256-bit hash or key that uniquely describes each block. This is an important security feature of blockchain technologies that is used for the private keys you receive that identify you as the owner of cryptocurrency.

A 256-bit private key has 115,792,089,237,316,195,423,570,985,008,687,907,853,269,984,665,640,564,039,457,584,007,913,129,639,936 possible combinations. It's a really, really big number. Even the fastest of today's classical computers would take millions of years to try all possible combinations to "hack" your private key.

But, if we build a quantum computer that could work with 256 qubits simultaneously, a hacker could conceivably try all possible combinations <u>at once</u>. Yikes!

Are We Going to Get There?

The answer is most certainly yes. We are just now beginning to see great advances in the quantum computing space, so it's going to take a while, but we will get there.

For example, Chinese startup company SpinQ offers a desktop quantum computer for $43,000 called Gemini. It also offers Gemini Mini, a scaled-down version, for $8,900. Both have a minimum configuration of 2 qubits but can operate as an 8-qubit simulator. [3]

Some of the big commercial competitors in this space include Honeywell, IBM, Google, Amazon, Microsoft, Rigetti, and Fujitsu. Most notable may be IBM. In 2021, IBM announced it had developed a quantum chip capable of working with 127 qubits. In 2022, it announced a quantum chip capable of working with 433 qubits. And more recently, IBM announced the first ever quantum chip capable of working with 1,000 qubits, 1,121 to be exact. [4]

Government initiatives/players in this space include:

- National University of Defense Technology, China (government security applications)
- National Supercomputer Center in Wuxi, China (climate science)
- Lawrence Livermore National Laboratory, United States (U.S. National Nuclear Security Administration modeling and simulations)
- U.S. Department of Energy, United States (energy efficiency, genetic applications)
- RIKEN Center for Computational Science (location of Fugaku), Japan (drug discovery, personalized medicine, climate forecasting, clean energy development)

Amazon has announced quantum-computing-as-a-service through Amazon Web Services (AWS). Which just happens to be a great segue into the next section on cloud computing.

\\\ CLOUD COMPUTING /// (AS-A-SERVICE)

(Cloud computing has been around for upwards of 30 years. So, it's definitely not a technology invented during the 4th industrial revolution. But the notion of "cloud" storage and processing are central to the success of innovations in the 4th industrial revolution.)

In the simplest of terms, ***cloud computing*** is the storing, accessing, and potentially processing of information and software on the Internet instead of your computer's hard drive. You're probably already "doing" cloud computing. If you use Dropbox for storing and sharing files, you're doing cloud computing. If you use iCloud and/or Google Drive, you're doing cloud computing.

Even things like Spotify are examples of cloud computing. With Spotify you can stream music from the Internet without having to store the songs on a device. (You can, of course, download your playlists from Spotify.) Amazon Prime is also an example of cloud computing. While you can download your video purchases from Amazon Prime for offline viewing, you don't have to. You can leave them in Amazon's "cloud," so to speak.

Cloud computing has become synonymous with "as-a-service." As-a-service is the opposite of ownership. Again, think about Spotify. You can access all the music you want as-a-service without paying for the actual ownership of the music. As long as you pay your monthly subscription, you can listen to all the music you want. Netflix is the same. Pay a monthly fee and watch all the movies you want. Of course, when you drop Netflix, you no longer have access to the movies because you didn't pay to own them.

The same is true for cloud computing. Instead of paying for ownership, you pay an as-a-service fee to have access.

Let's work through a simple example. Let's say you want to buy a flash drive to back up your computer. You have no intention of ever accessing the data on the flash drive unless your computer crashes or gets stolen. So, you've decided to buy a 2TB (2 terabyte) flash drive. That will cost you about $40. Small price to pay for back up, right? Indeed, but it can be even less in price if you use Amazon Web Services (AWS) and its S3 Glacier Deep Archive storage-as-a-service in the cloud. Using that service, 2TBs of backup data will cost you about $1 per month.

So, the equivalency is simple to determine. $40 now for a flash drive or $1 per month for 40 months (3 years and 4 months) spread out over that time period.

Of course, you can argue that flash drives are going to get cheaper in price over time (and that would be true), but that would imply that you would buy another flash drive whenever they dropped in price. Now, you've spent $40 on the original flash drive and perhaps another $30 18 months down the road when the price drops. Doesn't make sense.

As-a-service shows up in all aspects of our lives. People living in a big city like New York city often opt to not own a car but rather use transportation-as-a-service. If you own a car in New York City, you have all the typical costs of ownership… maintenance, parking, license plates, gas, insurance, etc. Or you can opt to pay for transportation - taxi cab, subway, Uber - when you need it and only when you need it.

Break-Even Analysis: The Business Decision of Cloud Computing

The business decision of cloud computing is simple and no different than your personal decision of transportation-as-a-service. Do you want to handle all the maintenance and other costs associated with your car, or do you want someone else to do so? Do you want to constantly pay for the "latest and greatest," or do you want to have as-a-service access to today's best? And, simply put, is it cheaper to buy and own or to "rent" what you need when you need it?

As you learn more about cloud computing from a business point of view, you'll see these 3 terms:

- ***Infrastructure-as-a-service (IaaS)*** - basic technology needs like virtual storage and servers.
- ***Platform-as-a-service (PaaS)*** - a step up from IaaS that also includes development environments and tools. For example, you may have applications you build for Windows operating systems (Microsoft), but you also want to build in the UNIX space.
- ***Software-as-a-service (SaaS)*** - the use of application software you need but don't want to pay for the ownership and associated costs.

Of the three, SaaS is the dominant model, accounting for over half of all cloud computing spending by businesses. [5] A good example is Salesforce. It was the first big player in the SaaS model for customer relationship management (CRM) software.

Cloud Providers

There are many providers in the cloud computing space. If you're interested in looking at any of them, consider the list below. [6]

- Amazon Web Services (AWS)
- Microsoft Azure
- Google Cloud
- Alibaba Cloud
- IBM Cloud
- DigitalOcean Cloud
- Oracle
- Salesforce
- ServiceNow
- Tencent Cloud
- Huawei Cloud
- Dell Technologies Cloud

Again, there are many.

\\\ NANOTECHNOLOGY: /// REALLY, REALLY SMALL

Okay, this is going to be a very short and sweet discussion. It's really challenging to provide a concise, yet approachable, definition of nanotechnology, so let's just chat for a bit.

Nanotechnology is all about extremely small things and using and manipulating those extremely small things to manufacture larger things. Incidentally, these "larger" things are also extremely small. When you think of nanotechnology, think of it as 3D printing on tiny steroids, lots of tiny steroids. Recall that subtractive manufacturing is taking some sort of material and whittling, slicing, and compressing it to make something smaller. We talked about trees before and making 2x4s, matches, and toothpicks from them. That's subtractive manufacturing.

Additive manufacturing is about building from the bottom up. In 3D printing, you manufacture something by continually adding layer upon layer of material until your product is complete.

The same is true for nanotechnology, but the "layer" becomes extremely small.

Nano stands for one-billionth. Not one billion, but rather one-billionth. So, 1 nanometer is one-billionth of a meter. There are just over 25 million nanometers in an inch.

- A human hair is about 100,000 nanometers in width.
- A sheet of paper is about 100,000 nanometers thick.
- Your fingernails grow about 1 nanometer per second. (How often do you have to cut your fingernails?)
- Your hair also grows about 1 nanometer per second. So, your hair grows about 1 inch in 25 million seconds or roughly 298 days. (That's, of course, the average. Yours may grow slower or faster.)
- The popular comparative scale: If a marble were a nanometer, then one meter would be the size of the earth. (Wrap your brain around that one.)

Hopefully, that gives you some sort of perspective for the scale of nanotechnology.

Formally, ***nanotechnology*** is the manipulation of material with at least one dimension of size between 1 and 100 nanometers. That's really, really, really small. Something that is 100 nanometers in width is 1,000 times smaller than a human hair. It takes unbelievably complex and expensive equipment to manufacture things on that scale of size.

But the possibilities of nanotechnology can go even smaller. Think about being able to build things on a molecule-by-molecule basis, or atom-by-atom basis. In case you've forgotten, atoms are single neutral particles. Molecules are neutral particles made up of two or more atoms bonded together. (That 7th grade general science class sure is coming in handy now, isn't it?)

It's theoretically possible using nanotechnology to build a computer - RAM, CPU, battery power supply, etc. - that would be about the width of a human hair.

Perhaps the most far-reaching and beneficial innovations of nanotechnology will be in the field of medicine. For example, researchers are developing nanoparticles to deliver drugs, heat, light, and other substances to specific types of cells. The nanoparticles are engineered in such a way that they are only attracted to certain types of diseased cells, such as cancer cells. In doing so, the nanoparticles would direct treatment only to the cancer cells, without damaging any of the surrounding healthy cells. [7]

Other researchers are working on building nanorobots that could be programmed to repair diseased cells. These nanorobots would emulate the antibodies of our natural healing process. [8]

Companies and Initiatives to Watch

- Applied Materials
- Avano
- CENmat
- CMC Materials
- DuPont de Nemours
- NanoScientifica
- Peak Nano
- Taiwan Semiconductor
- Tandem Nano
- Thermo Fisher Scientific

If you're really interested in nanotechnology, we would encourage you to explore the on-going research into the application of nanotechnology in medicine. Truly, truly fascinating.

ABOUT THE AUTHORS

Stephen Haag is a life-long techie, and now happily retired in Florida after teaching at the university level for 40+ years and authoring 50 books. His journey started in Borger, Texas at Frank Phillips Junior College and ended in Denver, Colorado at the University of Denver. While at DU, Stephen was Chair of the Department of Information Technology and Electronic Commerce, Director of Assurance of Learning, Director of Entrepreneurship, Director of the MBA, and Associate Dean of Graduate Programs. Stephen has four beautiful children (Darian, Trevor, Katrina, and Alexis). He is very fortunate to be married to Pam, the most wonderful and supportive soul mate ever.

Darian Haag is the oldest *(an important distinction)* son of Stephen Haag. Originally from Denver, Colorado, Darian attended Villanova University where he was a Finance and Management double major. It was at Villanova where Darian started his consulting career. Transitioning to New York City as his home base, Darian has worked at two different Big 4 firms, deepening his consulting and technology experience across a number of industries. He hopes to follow in his father's footsteps by driving technology innovation and enablement at businesses across the globe. Outside the office, Darian enjoys exercising, reading, traveling, and seeing his favorite artists at different venues around the country.

Trevor Haag is the best looking (an even more important distinction) son of Stephen Haag. Trevor attended the University of Denver where he received a Bachelor's degree in Finance and a Master's in Applied Quantitative Finance. Trevor's passions stem from his friends and video games, and he considers himself a jack of all trades and master of none.

ACKNOWLEDGEMENTS

Stephen Haag
The list is quite long. I wish I could name everyone who helped me get to this point in my life and my career. My mother and father (Nana and Popeye) have always been my foundation, my support, my biggest fans. My family, especially Pam, have loved and supported me regardless of my endeavors. My two co-authors, Darian and Trevor, are my sons. I've enjoyed writing with them and hope that they will continue. My two daughters, Katrina and Alexis, have always made me smile.

Darian Haag
This list will forever be growing, but there is a core group of people I want to thank. First and foremost is my family, grandparents, parents & siblings. None of this is possible without them. Next are all the friends I've made throughout the years. From the guys in Denver to everyone in New York City, thank you for the continued support you have for me. Finally, I want to acknowledge my mentors. From former coaches to professors to bosses, y'all's leadership and guidance have been vital to my growth both as a human and as a professional. Thanks to everyone along the way who helped make this possible.

Trevor Haag
I would like to thank my family, my father and mother Steve and Pam, my brother Darian, and my two younger sisters, Katrina and Alexis. I'd also like to thank my mentors, teachers, bosses, and friends who have shaped me into who I am today. Finally, thank you to my better half, Em.

Artwork
We will forever be thankful for our life-long friendships with Anton Prikhodko and Mykhailo Uvarov, both in Ukraine. Anton developed much of the artwork for the book, and Mykhailo crafted the cover design. If you need high-quality work, you can find both Anton and Mykailo on Upwork.

We also used VectorStock to acquire some of the artwork. Those pieces and their designers are listed on the next page.

Past/future p. 1 designed by ZdenekSasek (image #19731922)
IoT p. 13 designed by infadel (image #22721799)
Blockchain p. 26 designed by Anatoir (image #46435556)
Bitcoin p. 30 designed by Cowpland (image #15871239)
Artificial intelligence p. 51 designed by sumkinn (image #28838707)
Extended reality p. 75 designed by Valenty (image #19470008)
3D printing p. 91 designed by bsd_studio (image #39663766)
Autonomous vehicle p. 107 designed by Seamartini (image #43845597)
Drone p. 121 designed by bioraven (image #12166568)
Energy p. 131 designed by motorama (image #20529742)
Wind farm p. 138 designed by alexkava (image #47305361)
Senses p. 147 designed by nikolae (image #30058507)
Communications technology p. 161 designed by Pixelalex (image #11810904)
Infrastructure technology p. 169 designed by ForYoU13 (image #27825875)

NOTES

CHAPTER 1

1. Lee, Ahyoung & Wang, Xuan & Nguyen, H. & Ra, Ilkyeun. (2018). A Hybrid Software Defined Networking Architecture for Next-Generation IoTs. KSII Transactions on Internet and Information Systems. 12. 932-945. 10.3837/tiis.2018.02.024.
2. Watters, Ashley. "30 Internet of Things Stats & Facts for 2022." *CompTIA*, February 10 2022, https://connect.comptia.org/blog/internet-of-things-stats-facts.
3. "Industrial Internet of Things Market Trends." Grand View Research, https://www.grandviewresearch.com/industry-analysis/industrial-internet-of-things-iiot-market.
4. Trafton, Anne. "Sensors Woven into a Shirt Can Monitor Vital Signs." *MIT News | Massachusetts Institute of Technology*, MIT News Office, https://news.mit.edu/2020/sensors-monitor-vital-signs-0423.
5. Fisher, Tim. "What Is a Smart Bed?" *Lifewire*, February 3 2022, https://www.lifewire.com/smart-bed-4161313.
6. Sakharkar, Ashwini. "Printing Sensors Directly on Human Skin." *Tech Explorist*, October 9 2020, https://www.techexplorist.com/printing-sensors-directly-human-skin/35675/
7. Krebs, Brian. "Target Hackers Broke in via HVAC Company." *Krebs on Security*, February 5 2014, https://krebsonsecurity.com/2014/02/target-hackers-broke-in-via-hvac-company/
8. Williams, Robert. "Oral-B Connects AI-Powered Toothbrush to Mobile Apps for Personalized Tips." *Marketing Dive*, February 27 2019, https://www.marketingdive.com/news/oral-b-connects-ai-powered-toothbrush-to-mobile-apps-for-personalized-tips/549182/
9. Truong, Jessica. "How to Hack Self-Driving Cars: Vulnerabilities in Autonomous Vehicles." *HackerNoon*, July 13 2021, https://hackernoon.com/how-to-hack-self-driving-cars-vulnerabilities-in-autonomous-vehicles-jh3r37cz

CHAPTER 2

1. Prosser, Emmett. "Aaron Rodgers to take a part of his 2021 salary in Bitcoin, give $1 million of digital currency to fans." *USA Today*, November 1 2021, https://www.usatoday.com/story/sports/nfl/packers/2021/11/01/packers-aaron-rodgers-takes-portion-his-salary-bitcoin/6244838001/.
2. Foxley, William "Why the Navajo are mining Bitcoin." *Bitcoin Magazine*, November 4 2021, https://bitcoinmagazine.com/culture/why-the-navajo-are-mining-bitcoin.
3. Sigalos, MacKenzie. "Incoming Ney York mayor Eric Adams vows to take first three paychecks in bitcoin." *CNBC*, November 4 2021,

https://www.cnbc.com/2021/11/04/new-york-mayor-elect-eric-adams-to-take-first-3-paychecks-in-bitcoin.html.

4. NBC 6, "Miami to Distribute 'Bitcoin Yield' to Residents: Mayor Suarez." *NBC 6 South Florida*, November 11 2021, https://www.nbcmiami.com/news/local/miami-to-distribute-bitcoin-yield-to-residents-mayor-suarez/2618085/.
5. Bumbaca, Chris. "Trevor Lawrence partners with Blockfolio, will have signing bonus placed into cryptocurrency account." *USA Today*, April 26 2021, https://www.usatoday.com/story/sports/nfl/draft/2021/04/26/trevor-lawrence-jaguars-signing-bonus-cryptocurrency/7383149002/.
6. Zagorsky, Jay. "This country has just made bitcoin legal tender. Here's what it means." *World Economic Forum*, September 7 2021, https://www.weforum.org/agenda/2021/09/bitcoin-legal-tender-el-salvador-economics-finance/.
7. Elston, Thai-Binh. "China Is Doubling Down in Its Digital Currency," Foreign Policy Research Institute, June 2, 2023, https://www.fpri.org/article/2023/06/china-is-doubling-down-on-its-digital-currency/
8. "It costs 2₵ to make a penny and 7₵ to make a nickel, but CENTS Act could bring those costs down." *GovTrack Insider*, July 8 2019, https://govtrackinsider.com/it-costs-2-to-make-a-penny-and-7-to-make-a-nickel-but-cents-act-could-bring-those-costs-down-aa6aabfc9a8b.
9. Bharathan, Vipin. "US Postal Service Files A Patent For Voting System Combining Mail And A Blockchain." *Forbes*, September 20 2020, https://www.forbes.com/sites/vipinbharathan/2020/09/20/us-postal-service-files-a-patent-for--voting-system-combining-mail-and-a-blockchain/.
10. Henderson, James. "De Beers tracks 100 diamonds through supply chain using blockchain." *SupplyChain*, May 17 2020, https://supplychaindigital.com/technology/de-beers-tracks-100-diamonds-through-supply-chain-using-blockchain, Supple Chain Digital.
11. Vitasek Kate, Bayliss John, Owen Loudon, Srivastava Neeraj. "How Walmart Canada Uses Blockchain to Solve Supply-Chain Challenges." *Harvard Business Review*, January 5 2022, https://hbr.org/2022/01/how-walmart-canada-uses-blockchain-to-solve-supply-chain-challenges.
12. Daley, Sam. "19 Blockchain Companies Boosting the Real Estate Industry." *Builtin*, July 30 2021, https://builtin.com/blockchain/blockchain-real-estate-companies.
13. Daley, Sam. "How Using Blockchain in Healthcare is Reviving the Industry's Capabilities." *Builtin*, July 30 2021, https://builtin.com/blockchain/blockchain-healthcare-applications-companies.

14. Hydrogen. "5 Common Blockchain Applications in Financial Services." *Hydrogen*, December 12 2019, https://www.hydrogenplatform.com/blog/5-common-blockchain-applications-in-financial-services.
15. "The Unbanked." *Findex*, 2017, https://globalfindex.worldbank.org/sites/globalfindex/files/chapters/2017%20Findex%20full%20report_chapter2.pdf.
16. Mearian, Lucas. "10 top distributed apps (dApps) for blockchain." *ComputerWorld*, December 30 2019, https://www.computerworld.com/article/3510457/10-top-distributed-apps-dapps-for-blockchain.html.
17. Locke, Taylor. "What are DAOs? Here's what to know about the 'next big trend' in crypto." *CNBC*, October 25 2021, https://www.cnbc.com/2021/10/25/what-are-daos-what-to-know-about-the-next-big-trend-in-crypto.html.
18. Locke, Taylor. "What are DAOs? Here's what to know about the 'next big trend' in crypto." *CNBC*, October 25 2021, https://www.cnbc.com/2021/10/25/what-are-daos-what-to-know-about-the-next-big-trend-in-crypto.html.
19. "Top DAOs Projects in 2023: Full List," CFTE, March 7, 2023, https://blog.cfte.education/top-daos-projects-2023/.
20. Browne, Ryan. "More than $90 million in cryptocurrency stolen after a top Japanese exchange is hacked." *CNBC*, August 19 2021, https://www.cnbc.com/2021/08/19/liquid-cryptocurrency-exchange-hack.html.
21. George, Kevin. "The Largest Cryptocurrency Hacks So Far," *Investopedia*, December 2, 2023, https://www.investopedia.com/news/largest-cryptocurrency-hacks-so-far-year/.
22. Strydom, Lorien. "What Is the Best Stablecoin? 8 Top Stablecoins to Buy in 2024," *financer.com*, January 17, 2024, https://financer.com/us/blog/best-stablecoin-to-buy/.
23. Shewale, Rohit. "25 Most Expensive NFTs Ever Sold (2024)." demandsage, January 6, 2024, https://www.demandsage.com/most-expensive-nfts/.
24. Khatri, Amara. "Snoop Dogg To Sell 1000 Passes For Private Metaverse Party." *Bitcoin Insider*, September 24 2021, https://www.bitcoininsider.org/article/127968/snoop-dogg-sell-1000-nft-passes-private-metaverse-party.
25. Robinson, Randy. "Snoop Dogg's Metaverse Neighbors Paid $1.23 million for Digital Real Estate." *Merry Jane*, December 7 2021, https://merryjane.com/news/snoop-doggs-metaverse-neighbors-paid-dollar123-million-for-digital-real-estate.
26. "Best NFT Marketplaces for 2024." *The Ascent*, March 7, 2024, https://www.fool.com/the-ascent/cryptocurrency/nft-marketplaces/.

CHAPTER 3

1. Brown, Mike. "A Google Algorithm Was 100 Percent Sure That a Photo of a Cat Was Guacamole." *Inverse*, Inverse, June 20 2019, https://www.inverse.com/article/56914-a-google-algorithm-was-100-percent-sure-that-a-photo-of-a-cat-was-guacamole
2. "Sophia." *Hanson Robotics*, September 1 2020, https://www.hansonrobotics.com/sophia/
3. "Kismet, the Robot." *MIT Education*, http://www.ai.mit.edu/projects/sociable/baby-bits.html
4. "DeepMind Papers at ICML 2018." Google DeepMind, July 9, 2018, https://deepmind.google/discover/blog/deepmind-papers-at-icml-2018/
5. Malaviya, Sandip. "Top 10 AI Development Tools, Frameworks for 2021." *Samarpan Infotech*, Samarpan Infotech, September 29 2021, https://www.samarpaninfotech.com/blog/best-ai-development-tools-frameworks/
6. Daley, Sam. "28 Examples of Artificial Intelligence Shaking up Business as Usual." *Built In*, August 9 2021, https://builtin.com/artificial-intelligence/examples-ai-in-industry
7. Hagan, Shelly. "More Robots Mean 120 Million Workers Need to Be Retrained." *Bloomberg*, Bloomberg, September 5 2019, https://www.bloomberg.com/news/articles/2019-09-06/robots-displacing-jobs-means-120-million-workers-need-retraining

CHAPTER 4

1. Grannell, Craig. "Best Iphone AR Apps and Games in 2021." *Tom's Guide*, April 14 2021, https://www.tomsguide.com/round-up/best-iphone-ar-apps
2. Zoria, Sophie. *Enterprise AR: 7 Real-World Use Cases for 2021*. *AR/VR Journey: Augmented & Virtual Reality Magazine*, March 12 2021, https://arvrjourney.com/enterprise-at-7-real-world-use-cases-for-2021-81ea0319b8e5?gi=f8f1ab935e75
3. Saleem, Hania. "10 Best Tools for VR Development in 2023," *Talentverse*, January 30, 2023, https://www.talentverse.co/blog/best-tools-for-vr-development
4. "Virtual Reality Development Software Reviews and Rankings, *Gartner*, https://www.gartner.com/reviews/market/virtual-reality-development-software
5. "10 Business Applications of Virtual Reality (VR) Technology." *Sendian Creations*, Sendian Creations, February 8 2021, https://www.sendiancreations.com/top-vr-business-applications
6. Reid, Johnny. "Your Guide to Mixed Reality Headsets in the U.S.," *goHere AR*, Decemver 23, 2023,

https://www.mixyourreality.com/insights/mixed-reality-headsets-available-in-the-us
7. Snider, Mike, and Brett Molina. *USA Today*, November 11 2021, https://www.usatoday.com/story/tech/2021/11/10/metaverse-what-is-it-explained-facebook-microsoft-meta-vr/6337635001.
8. Kafka, Peter. "Facebook Is Quietly Buying up the Metaverse." *Vox*, November 11 2021, https://www.vox.com/recode/22776461/facebook-meta-metaverse-monopoly
9. Warren, Tom. "Microsoft Teams Enters the Metaverse Race with 3D Avatars and Immersive Meetings." *The Verge*, November 2 2021, https://www.theverge.com/2021/11/2/22758974/microsoft-teams-metaverse-mesh-3d-avatars-meetings-features?scrolla=5eb6d68b7fedc32c19ef33b4
10. Thurman, Andrew. "Barbados to Become First Sovereign Nation with an Embassy in the Metaverse." *Yahoo!*, November 15 2021, https://www.yahoo.com/now/barbados-become-first-sovereign-nation-110000022.html

CHAPTER 5

1. "3D Print-Knit." *Ministry of Supply*, https://www.ministryofsupply.com/technologies/3d-print-knit
2. Scott, Clare. "Julia Daviy Uses 3D Printing to Create Beautiful Biodegradable Fashion." *3DPrint.Com*, May 7 2018, https://3dprint.com/212640/julia-daviy-3d-printing
3. *VIP TIE - Sustainable Luxury Accessories - Dubai, Milan*, https://www.viptie3d.com
4. Fox News. "Chinese Company Builds First 3D-Printed Apartment Building, Mansion." *Fox News*, FOX News Network, January 20 2015, https://www.foxnews.com/tech/chinese-company-builds-first-3d-printed-apartment-building-mansion
5. Rogers, James. "Marines 3D-Print Concrete Barracks in Just 40 Hours." *Fox News*, FOX News Network, August 29 2018, https://www.foxnews.com/tech/marines-3d-print-concrete-barracks-in-just-40-hours
6. Howarth, Dan. "Icon Completes 'First 3D-Printed Homes for Sale in the US.'" *Dezeen*, August 31 2021, https://www.dezeen.com/2021/08/31/east-17th-street-residences-3d-printed-homes-icon-austin
7. Peters, Adele. "This Wild-Looking House Is Made out of Dirt by a Giant 3D Printer." *Fast Company*, April 9 2021, https://www.fastcompany.com/90619146/this-wild-looking-house-is-made-out-of-dirt-by-a-giant-3d-printer

8. "7 Exciting 3D Printed Food Projects Changing How We Eat Forever." *3DSourced*, August 19 2021, https://www.3dsourced.com/guides/3d-printed-food
9. Corolo, Lucas. "3D Printed Food: All You Need to Know in 2021." *All3DP*, November 22 2021, https://all3dp.com/2/3d-printed-food-3d-printing-food
10. Varkey, M., Visscher, D.O., van Zuijlen, P.P.M. *et al.* Skin bioprinting: the future of burn wound reconstruction?. *Burn Trauma* **7,** 4 (2019). https://doi.org/10.1186/s41038-019-0142-7
11. *Collplant*, https://collplant.com
12. *Dimension Inx*, https://www.dimensioninx.com
13. *3DBio Therapeutics*, https:/3dbiocorp.com
14. *Frontier Bio*, https://www.frontierbio.com
15. *Printivo*, https://www.printivo.eu
16. Affordable Dentures & Implants, https://www.affordabledentures.com/our-services/dentures/realfit-3d-dentures
17. *Enabling The Future*, https://enablingthefuture.org
18. Gonzalez, Carlos. *The Future of 3D Printing on Human Skin*. Machine Design, May 3 2018, https://www.machinedesign.com/3d-printing-cad/article/21836698/the-future-of-3d-printing-on-human-skin
19. D., Jamie. "Top 12 Best Websites to Download Free STL Files." *3Dnatives*, March 17 2020, https://www.3dnatives.com/en/top-10-websites-stl-files-161120174
20. "Software for 3D Printing." *3D Printing*, https://3dprinting.com/software
21. "Software for 3D Printing." *3D Printing*, https://3dprinting.com/software
22. "Paper and Paperboard: Material-Specific Data ." *EPA*, Environmental Protection Agency, https://www.epa.gov/facts-and-figures-about-materials-waste-and-recycling/paper-and-paperboard-material-specific-data

CHAPTER 6

1. U.S. Department of Transportation. National Highway Traffic Safety Administration. *Critical Reasons for Crashes Investigated in the National Motor Vehicle Crash Causation Survey*, 2015. *DOT HS 812 115.* https://crashstats.nhtsa.dot.gov/Api/Public/ViewPublication/812115.
2. U.S. General Services Administration. Office of Motor Vehicle Management. *Crashes Are No Accident.* https://drivethru.gsa.gov/DRIVERSAFETY/DistractedDrivingPosterA.pdf.
3. "NHTSA Estimates for 2022 Show Roadway Fatalities Remain Flat After Two Years of Dramatic Increases," *NHTSA*, April 20, 2023, https://www.nhtsa.gov/press-releases/traffic-crash-death-estimates-2022.

4. U.S. General Services Administration. Office of Motor Vehicle Management. *Crashes Are No Accident.* https://drivethru.gsa.gov/DRIVERSAFETY/DistractedDrivingPosterA.pdf
5. "Drunk Driving," NHTSA, https://www.nhtsa.gov/risky-driving/drunk-driving.
6. "SAE Levels of Driving Automation™ Refined for Clarity and International Audience." *SAE International*, May 3 2021, https://www.sae.org/blog/sae-j3016-update.

CHAPTER 7

1. "The 13 Best Drones You Can Buy," *Popular Mechanics*, https://www.popularmechanics.com/technology/gadgets/g32209219/top-drones/.
2. "Using Drones to Deliver Blood in Rwanda." *BBC NEWS*, March 19 2019, https://www.bbc.com/news/av/business-47631709.
3. "Kroger and Drone Express Partner to Provide Grocery Delivery by Drone." *The Kroger Co.*, March 5 2021, https://ir.kroger.com/CorporateProfile/press-releases/press-release/2021/Kroger-and-Drone-Express-Partner-to-Provide-Grocery-Delivery-by-Drone/default.aspx.
4. Peters, Jay. "UPS Just Won FAA Approval to Fly as Many Delivery Drones as It Wants." *The Verge*, October 1 2019, https://www.theverge.com/2019/10/1/20893655/ups-faa-approval-delivery-drones-airline-amazon-air-uber-eats-alphabet-wing.
5. "Wing Drone Deliveries Take Flight with Fedex®." *FedEx*, October 18 2019, https://www.fedex.com/en-us/sustainability/wing-drones-transport-fedex-deliveries-directly-to-homes.html.
6. "First Prime Air Delivery." *Amazon*, https://www.amazon.com/Amazon-Prime-Air/b?node=8037720011.
7. "Drones for Police, Fire and Public Safety." *Blue Skies Drones*, https://www.blueskiesdronerental.com/first-responders.
8. "Drones for Agriculture." *Precision Hawk*, https://www.precisionhawk.com/agriculture/drones.
9. French, Sally. "What Real Estate Agents Should Know about Drones for Aerial Photography." *The Drone Girl*, October 5 2021, https://www.thedronegirl.com/2021/05/21/real-estate-agents-using-drones.
10. "CityAirbus NextGen." *Airbus*, https://www.airbus.com/en/innovation/zero-emission/urban-air-mobility/cityairbus-nextgen.
11. Vijayenthiran, Viknesh. "Toyota-Backed Flying Taxi Prototype Takes to the Skies." *Motor Authority*, September 1 2020,

https://www.motorauthority.com/news/1129452_toyota-backed-flying-taxi-prototype-takes-to-the-skies.
12. Vijayenthiran, Viknesh. "Toyota-Backed Flying Taxi Startup Joby Aviation Goes Public via SPAC Deal." *Motor Authority*, August 13 2021, https://www.motorauthority.com/news/1131408_toyota-backed-flying-taxi-startup-joby-aviation-goes-public-via-spac-deal.
13. Vijayenthiran, Viknesh. "Hyundai Launches Supernal Flying Taxi Division, Promises First Commercial Flight by 2028." *Motor Authority*, November 10 2021, https://www.motorauthority.com/news/1132611_hyundai-launches-supernal-flying-taxi-division-promises-first-commercial-flight-by-2028.
14. McFarland, Matt. "Boeing's First Autonomous Air Taxi Flight Ends in Fewer than 60 Seconds." *CNN*, January 23 2019, https://www.cnn.com/2019/01/23/tech/boeing-flying-car/index.html.
15. Trop, Jaclyn. "Yes, Flying Cars Are Coming. Here Are 7 That Are Hitting the Skies Soon, Robb Report, April 18, 2023, https://robbreport.com/motors/aviation/gallery/7-flying-cars-change-air-transport-1234831349/.
16. Tuohy, Jennifer Pattison. "Amazon Is Now Accepting Your Applications for Its Home Surveillance Drone." *The Verge*, September 21 2021, https://www.theverge.com/2021/9/28/22692048/ring-always-home-cam-drone-amazon-price-release-date-specs.

CHAPTER 8

1. "U.S. energy facts explained," *U.S. Energy Information Administration*, April 2023, https://www.eia.gov/energyexplained/us-energy-facts/.
2. Cox, Chelsey, and Michelle Shen. "What Are the Effects of Climate Change Costing Consumers, on Average?" *USA Today*, Gannett Satellite Information Network, November 2 2021, https://www.usatoday.com/story/money/2021/11/02/what-climate-change-costing-average-american-per-month/8544943002.
3. "Just 1 Second of the Sun's Energy Output Would Power the US for 9,000,000 Years." *Wow Really*, https://www.wowreally.blog/2006/10/ust-1-second-of-suns-energy-output.html#:~:text=Just%201%20second%20of%20the,the%20US%20for%209%2C000%2C000%20years.
4. "Ivanpah," *Energy.gov*, https://www.energy.gov/lpo/ivanpah.
5. Bekele, Adisu, et al. "Large-Scale Solar Water Heating Systems Analysis in Ethiopia: A Case Study." *International Journal of Sustainable Energy*, vol. 32, no. 4, September 23 2011, pp. 207–228., https://doi.org/10.1080/14786451.2011.605951.
6. Chandler, David L. "Simple, Solar-Powered Water Desalination." *MIT News*, February 6 2020, https://news.mit.edu/2020/passive-solar-powered-water-desalination-0207.

7. *Household Water Treatment Options in Developing Countries: Solar Disinfection (SODIS)*. Centers for Disease Control and Prevention & US Aid, January 2008, https://web.archive.org/web/20080529090729/http://www.ehproject.org/PDF/ehkm/cdc-options_sodis.pdf.
8. Weise, Elizabeth. "This State Is Quickly Becoming America's Clean Energy Paradise. Here's How It's Happening," USA Today, January 27, 2024, https://www.usatoday.com/story/news/nation/2024/01/27/how-one-state-is-quickly-becoming-americas-clean-energy-paradise/72252758007/.
9. "Hydrogen Explained - Use of Hydrogen." U.S. Energy Information Administration (EIA), January 20 2022, https://www.eia.gov/energyexplained/hydrogen/use-of-hydrogen.php.
10. D'Allegro, Joe. "Elon Musk Says the Tech Is 'Mind-Bogglingly Stupid,' but Hydrogen Cars May Yet Threaten Tesla." *CNBC*, February 24 2019, https://www.cnbc.com/2019/02/21/musk-calls-hydrogen-fuel-cells-stupid-but-tech-may-threaten-tesla.html.
11. Schwartz, Ariel. "A French Sidewalk Lets You Power the Streetlights With Your Feet." *Fast Company*, April 15 2010, https://www.fastcompany.com/1617178/french-sidewalk-lets-you-power-streetlights-your-feet.
12. Ross, Philip. "Good Vibrations? California to Test Using Road Rumbles as a Power Source," IEEE Spectrum, April 9, 2017, https://spectrum.ieee.org/good-vibrations-california-to-test-road-vibrations-as-a-power-source.
13. *Solar Marketplace Intel Report (Report) (13th ed.), EnergySage, August 2021.*
14. Cohen, Ariel. "What Batteries Will Power the Future?" *Forbes*, Forbes Magazine, February 11 2021, https://www.forbes.com/sites/arielcohen/2021/02/11/what-batteries-will-power-the-future/?sh=53bb21bb41c0.

CHAPTER 9

1. "Automatic Speech Recognition (ASR) Systems Compared." *Medium*, Sciforce, July 7 2021, https://medium.com/sciforce/automatic-speech-recognition-asr-systems-compared-6ad5e54fd65f.
2. "GnoSys App Translates Sign Language into Speech in Real Time Using the Power of AI." *Newz Hook*, October 28 2018, https://newzhook.com/story/20387.

CHAPTER 10

1. "What Are the Differences between 2G, 3G, 4G LTE, and 5G Networks?" *RantCell*, https://rantcell.com/comparison-of-2g-3g-4g-5g.html.

CHAPTER 11

1. Metz, Cade. "Google Claims a Quantum Breakthrough That Could Change Computing." *The New York Times*, October 23 2019, https://www.nytimes.com/2019/10/23/technology/quantum-computing-google.html.
2. Outeiral, Carlos, et al. WIREs, 2020, *The Prospects of Quantum Computing in Computational Molecular Biology*, https://doi.org/10.1002/wcms.1481.2.
3. Brassell, Jack. "SpinQ Is Selling Quantum Computers to Consumers," *BEYONDGAMES.biz*, December 21, 2022, https://www.beyondgames.biz/29903/spinq-is-selling-quantum-computers-to-consumers/.
4. Castelvecchi, Davide. "IBM Releases First-Ever 1,000-Qubit Quantum Chip," *Scientific American*, December 5, 2023, https://www.scientificamerican.com/article/ibm-releases-first-ever-1-000-qubit-quantum-chip/.
5. Ranger, Steve. "What Is Cloud Computing? Everything You Need to Know about the Cloud Explained." *ZDNet*, December 13 2018, https://www.zdnet.com/article/what-is-cloud-computing-everything-you-need-to-know-about-the-cloud.
6. Slingerland, Cody. "11 Top Cloud Service Providers Globally in 2024," *CloudZero*, December 15, 2023, https://www.cloudzero.com/blog/cloud-service-providers/.
7. "Nanotechnology and Nanoparticles in Drug Delivery." *UnderstandingNano*, https://www.understandingnano.com/nanotechnology-drug-delivery.html.
8. Freitas, Robert A. "The Ideal Gene Delivery Vector: Chromallocytes, Cell Repair Nanorobots for Chromosome Replacement Therapy." *Journal of Evolution and Technology*, vol. 16, no. 1, June 2007, pp. 1–97. *ISSN 1541-0099.*

INDEX/DEFINITIONS

2G, 166
360 camera (omnidirectional camera), 152
3D printing, 92
3D reprinting, 104
3G, 166
4G, 166
5G, 166
6G, 166

A
Actuator, 149
Additive manufacturing, 92
Alkaline battery, 134
Artificial general intelligence (AGI), 56
Artificial intelligence (AI), 52
Artificial narrow intelligence (ANI, narrow AI), 53
Artificial neural network (ANN), 62
Artificial superintelligence (ASI), 59
Augmented reality (AR), 77
Autonomous drone, 122
Autonomous vehicle, 108

B
Battery, 143
Barcode (bar code), 159
Biomass, 140
Biometrics, 154
Biomimicry, 68
Bioprinting (3D bioprinting), 97
Bit, 170
Block hash or key, 36
Blockchain, 34
Blockchain node (node), 34
Bluetooth, 162
British Thermal Unit (BTU), 133
Broadband, 165
Broadband cellular network, 165
Build area, 103

C
Cellular network, 165
Central bank digital currency (CBDC), 30
Climate change, 134
Cloud computing, 177
Cold cryptocurrency wallet (cold wallet, cold storage, hardware-based cryptocurrency wallet), 42
Computer vision, 151
Concentrated solar power (CSP), 137
Convolutional neural network (CNN), 67
Cryptocurrency (crypto), 30
Cryptocurrency exchange, 41
Cryptocurrency mining (mining), 43
Cryptocurrency wallet, 41

D
Decentralized app (Dapp), 39
Decentralized autonomous organization (DAO), 39
Decentralized finance (DeFi), 38
Deep learning, 64
Digital age, 7
Digital immigrant, 7
Digital Light Processing (DLP), 102
Digital native, 7
Digital revolution, 7
Digital twin, 157
Distributed ledger technology (DLT), 33
Driver-assistance technology, 110
Drone, 122

E
Edge computing, 23
Electric vehicle battery (EVB), 145
EMV (Europay, Mastercard, Visa), 164
Energy, 132
Energy harvesting, 136
Environmental, social, and corporate governance (ESG), 146
eVTOL (electric vertical take-off and landing), 128
Expert system, 54
Extended reality (XR), 76
Eye tracking, 155

F
Fiat money, 29
Filament, 98
Fintech, 38
Fused deposition modeling (FDM), 102

G
G-code, 101
Generative artificial intelligence (generative AI, GenAI), 59
Genetic algorithm, 67
Geothermal power, 139
Gesture recognition, 154
Global warming, 134
Greenhouse gas (GHG or GhG), 134
Greenhouse effect, 134

H
Hand and finger tracking, 155
Haptic technology, 156
Hearing technology, 150
Hologram, 89
Hot cryptocurrency wallet (hot wallet, hot storage), 41
Hydrogen fuel, 141
Hydropower (water power), 139

I
Industrial Internet of Things (IIoT), 17
Information age, 7
Infrastructure-as-a-service (IaaS), 179
Internet of Things (IoT), 15
Internet of All Things (IoAT), 18
Internet of Computer Things (IoCT), 15
Internet of Electronic Things (IoET), 16

L
Large language model (LLM), 60
Latency, 166
Legal tender, 29
Lidar, 152
LiFi (light fidelity, Li-Fi), 165
Limited-memory artificial intelligence, 56
Lithium-ion battery, 144

M
Machine learning, 64
Matter, 17
Metaverse, 85
Microwave, 167
Mixed reality (MR), 82
Mobile solar (MoSo), 137
Motion capture (mocap, mo-cap), 154

N
Nano, 180
Nanotechnology, 180
Near-field communication (NFC), 164
Neuroevolution, 68
Non-fungible token (NFT), 47

O
Optical character recognition (OCR), 153

P
Paris Agreement (Paris Accords, Paris Climate Agreement, or Paris Climate Accords), 142
Photovoltaic system (PVC), 137
Piezoelectricity, 142
Platform-as-a-service (PaaS), 179
Primary (non-rechargeable) battery, 143
Private key, 40
Prompt engineering, 60
Proof-of-stake mining, 44
Proof-of-work mining, 43
Public key, 40
Public key cryptography (asymmetric encryption), 40

Q
QR (quick response) code, 159
Quantum computer (quantum computing), 172
Qubit, 172

R
Radar, 152
Radio-frequency identification (RFID), 163
Reactive artificial intelligence, 54
Renewable energy, 134
Resin, 98

S
Satellite, 167
Secondary (rechargeable) battery, 134
Self-aware artificial intelligence, 59
Sensor, 21
SLAM (simultaneous localization and mapping), 154
Slicing software, 101
Smart city, 17
Smart contract, 38
Smart grid, 17
Smart home (home automation, domotics), 16
Smart mirror, 88
Smelling technology (digital scent technology, olfactory technology), 157
Software-as-a-service (SaaS), 179
Solar power, 136
Solid-state battery, 145
Sonar, 152
Speech recognition (automatic speech recognition, ASR), 150
Stablecoin, 43
STL file (Standard Tessellation Language, StereoLithography File), 99
Subtractive manufacturing, 92
Supervised learning, 66

T
Tethering, 168
Theory of mind artificial intelligence, 58
Transducer, 149

U
Un-banked, 38
Universal product code (UPC), 159
Unmanned aerial vehicle (UAV), 122
Unsupervised learning, 66

V
Virtual reality (VR), 79

W
Web 1.0, 71
Web 2.0, 71
Web 3.0, 71
WiFi (wireless fidelity), 164
Wind farm, 138
Wind power, 138

Made in the USA
Las Vegas, NV
13 September 2024

95236282R00115